CURRENT PERSPECTIVE TO PREDICT ACTUAL EVAPOTRANSPIRATION

Edited by **Daniel Bucur**

Current Perspective to Predict Actual Evapotranspiration
http://dx.doi.org/10.5772/65621
Edited by Daniel Bucur

Contributors

Regis Barille, Antonin Poisson, Angel Fernandez, Benoit Chorro, Dario Gomez, Hsin-Fu Yeh, Carlos Cortés-Martínez, Jaime Ruiz-Vega, Gabino Martínez-Gutiérrez, Wonsook Ha, Abraham Springer, Frances O'Donnell, Thomas Kolb, Gilberto Javier Fochesatto, Watcharee Ruairuen

Published by InTech
Janeza Trdine 9, 51000 Rijeka, Croatia

Notice

Publishing Process Manager Edi Lipovic
Technical Editor SPi Global
Cover InTech Design team

First published May, 2017
Printed in Croatia
Legal deposit, Croatia: National and University Library in Zagreb

Additional hard copies can be obtained from orders@intechopen.com

Current Perspective to Predict Actual Evapotranspiration, Edited by Daniel Bucur
p. cm.
Print ISBN 978-953-51-3173-1
Online ISBN 978-953-51-3174-8

For EU product safety concerns:
IN TECH d.o.o., Prolaz Marije Krucifikse Kozulić 3, 51000 Rijeka, Croatia,
info@intechopen.com or visit our website at intechopen.com.

Meet the editor

Daniel Bucur is currently a professor for Land Improvement at the University of Applied Life Sciences and Environment in Iasi, Romania. He completed his doctorate degree at the Technical University of Iasi in 1998.
His major research areas include water excess removal, irrigation, soil erosion control, climate changes, and sustainable land management. In recent years, he has been in charge of many national and international research projects, including *improving soil and water resource management in the Moldavian Plain, soil erosion and conservation measures, effect of sewage sludge application on quality indices of soil vulnerable to degradation, sustainable development of soil resources from the areas with drainage works*, and *impact of the hydroclimatic and pedo-geomorphological risks on the environment in small catchment*. He has published more than 150 papers in reviewed journals, 6 book chapters, and 11 books apart from more than 35 unreviewed papers and reports.

Contents

Preface IX

Chapter 1 **Comparison of Evapotranspiration Methods Under Limited Data 1**
Hsin-Fu Yeh

Chapter 2 **Assessment and Prediction of Evapotranspiration Based on Scintillometry and Meteorological Datasets 25**
Antonin Poisson, Angel Fernandez, Dario G. Gomez, Régis Barillé and Benoit Chorro

Chapter 3 **Sensitivity of Evapotranspiration Models to Onsite and Offsite Meteorological Data for a Ponderosa Pine Forest 47**
Wonsook Ha, Abraham E. Springer, Frances C. O'Donnell and Thomas E. Kolb

Chapter 4 **Evapotranspiration in Northern Agro-Ecosystems: Numerical Simulation and Experimental Comparison 65**
Watcharee Ruairuen, Gilberto J. Fochesatto, Marco Bittelli, Elena B. Sparrow, Mingchu Zhang and William Schnabel

Chapter 5 **Moisture Evaporation from Granular Biopesticides Containing Quiescent Entomopathogenic Nematodes 85**
Carlos Inocencio Cortés-Martínez, Jaime Ruiz-Vega and Gabino Alberto Martínez-Gutiérrez

Preface

Water resources are highly influenced by the hydrologic cycle and play a role in the agriculture economic development. However, as it is shown by the intergovernmental panel on climate change report, the phenomenon of changing climate has set way to exacerbate an already serious situation of water supply for various users. In this context, the scientific investigations on the issue of sustainable use of water are timely and important. Improvement of water management involves the accurate estimation of consumptive uses. One of the techniques is the assessment of evapotranspiration, as a major component of the hydrologic cycle.

As is known, some of the meteorological parameters necessary to estimate evapotranspiration are frequently missing or hard to collect, and the maintenance of meteorological stations is costly. Therefore, the study presented in the first chapter aimed to estimate evapotranspiration using a limited number of meteorological parameters. With the internationally accepted Penman-Monteith method as the standard, the estimation formulas of radiation-based methods were compared with those of temperature-based methods in the hope of discovering a simple estimation formula to solve the issue of lacking or missing meteorological data.

In order to anticipate the necessary action for water management, researchers from Europe and South America have developed a model to forecast evapotranspiration based on an artificial neural network. The predicted output and final results are compared for two different input parameters of evapotranspiration calculated with scintillometry and micrometeorology data. The goal of the study is to find which input data can be reliable to obtain evapotranspiration forecast performances. Moreover, the optimal number of predicted days to obtain a correct final performance and the optimal number of input days of data to obtain a correct prediction are tested.

Given the same limitations—time and resources—data required to model evapotranspiration are not always available for a study site. But, North American specialists consider that off-site data from meteorological networks may be a suitable substitute. In their study, sensitivity to mixtures of on-site and off-site data inputs of three widely used models for calculating evapotranspiration for a ponderosa pine forest where evapotranspiration was measured previously by eddy covariance was investigated.

Northern latitudes have been identified as a region where global climate change will have earlier and stronger impacts than in other regions of the world. All these changes will potentially alter the exchange of surface energy, water, and carbon cycles in high latitude ecosystems and consequently the response at the regional level to the atmosphere system. Accordingly, an interesting study, based on current numerical model outputs and comparisons with field experimental observations, allowed to identify new challenges in northern agroecosystems.

Moisture evaporation from porous media is studied by its importance in drying of foods and building materials and biological products such as biopesticides. An exciting foray into the current state of knowledge in the frame of physics of moisture evaporation process from porous media is the subject of the last chapter. Further, the authors aim to establish theoretical support for designing biopesticides able to ensure efficient and effective fight against harmful insects in agricultural crops.

Prof. Dr. Daniel Bucur,
University of Applied Life Sciences and Environment in Iasi,
Romania

Chapter 1

Comparison of Evapotranspiration Methods Under Limited Data

Hsin-Fu Yeh

Additional information is available at the end of the chapter

http://dx.doi.org/10.5772/intechopen.68495

Abstract

A limited number of parameters or a single meteorological parameter was used in this study to estimate evapotranspiration. The main objectives of this study are as follows. (1) The Penman-Monteith method was used to estimate ET. The empirical formula published by the Food and Agriculture Organization (FAO) was applied via substitution to compare situations that were missing certain meteorological parameters. (2) Radiation-based methods and temperature-based methods were compared with the Penman-Monteith method to estimate ET and discuss their applicability in the study area. With Tainan Weather Station of Taiwan as the study area, this study selected the Penman-Monteith method as well as six other radiation-based estimation formulas: Makkink, Turc, Jensen-Haise, Priestley-Taylor, Doorenbos-Pruit, and Abtew methods. The other four temperature-based estimation formulas, namely, Thornthwaite, Blaney-Criddle, Hamon, and Linacre methods, were used to estimate ET and compare the differences and the results were compared with the Penman-Monteith method. The results showed that there was little effect on estimating ET using the Penman-Monteith method when the wind speed data was missing or insufficient. The Turc method was the best among the six radiation-based estimation formulas, while the Linacre method was the best temperature-based estimation formula. Generally speaking, radiation-based estimation formulas were more accurate than temperature-based estimation formulas.

Keywords: evapotranspiration, Penman-Monteith, radiation method, temperature method

1. Introduction

Evapotranspiration (ET) is a basic element of the hydrologic cycle as well as a key factor in water balance [1]. According to statistics, global average annual rainfall is around 973 mm, and about 64% of surface water is lost through ET [2]. Therefore, ET is considered to be an indispensable parameter in hydrologic studies, such as irrigation scheduling and management, crop water demand, and environmental impact assessment [3]. Hence, effective evaluation of ET is important for the management and planning of water resources. In previous studies, many formulas of empirical or physical methods have been used to estimate ET in various climatic conditions; examples include the Makkink method [4], Priestley-Taylor method [5], lysimeter method [6], and micro-meteorological observation method [7]. The empirical formula of the Penman-Monteith method released by the Food and Agriculture Organization (FAO) is the method most internationally used [8]. This method requires consideration of a variety of meteorological parameters, such as temperature, radiation, relative humidity, and wind speed. These data, however, are frequently missing or hard to collect, resulting in difficulties in estimation [9]. In particular, reliable meteorological data, such as radiation, relative humidity, and wind speed, are rather difficult to collect in some areas. In addition, the maintenance of meteorological stations requires substantial funding and the installation is complex.

Therefore, in previous studies, many scholars have used a limited number of parameters or a single meteorological parameter to easily estimate ET and simplify the estimation methods, which are classified into five major categories based on the required meteorological parameters: (1) water balance method, (2) mass transfer method, (3) mixing method, (4) radiation-based method, and (5) temperature-based method [10]. Except for the last two methods, the other three methods require a variety of meteorological parameters to estimate ET, thus causing obstacles in data collection and obtaining complete meteorological information. Furthermore, studies have found that the results of empirical methods should be compared with the Penman-Monteith method and released by FAO so as to carry out accurate estimation in each region [11].

In this study, a single meteorological parameter was applied, as well as the Penman-Monteith method, six radiation-based methods, and four temperature-based methods, to effectively estimate ET. The main objectives of this study are as follows: (1) when radiation, wind speed, and relative humidity data were missing, empirical formulas were used for substitution in the Penman-Monteith method to compare the estimation results; (2) the regional applicability of the radiation- and temperature-based methods were compared so as to make these methods more suitable for the study area.

2. Material and methods

This study mainly discussed the effective evaluation of ET using limited meteorological parameters. With the Penman-Monteith method as the standard for estimation, ET was calculated using substitution formulas when radiation, wind speed, or relative humidity data were missing in the Penman-Monteith method. Six radiation-based methods and four temperature-based methods were selected to discuss their applicability in the study area. In this study, mean bias error, root mean square error, and the Pearson-type goodness-of-fit index were used to analyze and investigate the differences among ET estimations using the empirical formulas of temperature and radiation methods. Meanwhile, this study strived to determine the method with a simpler empirical formula to address the difficulties caused by a shortage of meteorological parameter data.

2.1. Penman-Monteith method

Penman-Monteith method was recommended by the FAO in the 1998 FAO-56 report for the assessment of ET, and it is currently used internationally [12]. After years of study by domestic scholars, it is believed that the Penman-Monteith method is quite suitable in Taiwan [13–15]. Its formula can be expressed as follows:

$$ET = \frac{0.408\Delta(R_n - G) + \gamma\frac{900}{T+273}u_2(e_s - e_a)}{\Delta + \gamma(1 + 0.34u_2)} \tag{1}$$

In Eq. (1), ET represents evapotranspiration (mm d^{-1}); Δ represents the slope of air pressure curve (kPa °C^{-1}); T is the average temperature (°C); R_n is net radiation (MJ m^{-2} d^{-1}); G is the soil thermal flux (MJ m^{-2} d^{-1}); γ is the humidity constant (kPa °C^{-1}); u_2 is the wind speed measured at the height of 2 m (m s^{-1}); and (e_s−e_a) is the difference between saturated and actual vapor pressure (kPa). For field applications, Eq. (1) was calculated with monthly air temperature, humidity, radiant energy, wind speed, and other parameters [12].

When data of some meteorological parameters could not be obtained or were incomplete, for instance, radiation, relative humidity, and wind speed, a calculation was conducted using the following empirical formula:

1. When data of relative humidity could not be obtained or was incomplete:

$$e_a = 0.611\exp\left(\frac{17.27T_{min}}{T_{min}+237.3}\right) \tag{2}$$

In Eq. (2), T_{min} represents minimum temperature (°C).

2. When radiation data could not be obtained or was incomplete:

$$R_s = k_{Rs}\sqrt{(T_{max} - T_{min})}R_a \tag{3}$$

In Eq. (3), k_{Rs} is the empirical coefficient (k_{Rs} = 0.19); R_a is extraterrestrial solar radiation (MJ m^{-2} d^{-1}).

3. When data of wind speed could not be obtained or was incomplete:

When there is no record of wind speed in the evaluation area, the average Taiwan wind speed of 1.83 m s^{-1} was used, which was estimated with the data collected by 20 central meteorological observatories in Taiwan during 1990–2008 [15]. In addition, wind speed at a height of 2 m above the ground was primarily used in the estimation of wind speed. Provided that the measurement height was not 2 m, the following formula was applied:

$$u_2 = u_z \frac{4.87}{\ln(67.8z - 5.42)} \tag{4}$$

In Eq. (4), u_z is the wind speed measured at a meteorological station (m s^{-1}); z is the height of the anemometer above the ground (m).

2.2. Radiation-based methods

Priestley and Taylor [5] proposed that the estimation of ET could be explored from the perspective of energy conversion on the water surface. Evapotranspiration increased with an increase of radiation. Hence, radiation was taken as a vital meteorological parameter for ET assessment. Radiation-based methods were mainly based on the simplified principle of energy balance to estimate ET. Therefore, ET could be evaluated using a single meteorological parameter, and, in general, the form of radiation-based methods is as follows:

$$ET = \frac{C_r}{\lambda}(wR_s) \text{ or } ET = \frac{C_r}{\lambda}(wR_n) \tag{5}$$

λ represents the latent heat of evaporation (MJ kg^{-1}); C_r represents the generated empirical coefficient based on the relative humidity and wind speed; w is the generated empirical coefficient in accordance with temperature and latitude; R_s represents the amount of solar radiation (MJ m^{-2} d^{-1}); and R_n is the net radiation (W m^{-2} d^{-1}).

Six radiation-based methods that are used internationally to assess evapotranspiration were selected in this study, including Makkink [4], Turc [16], Jensen-Haise [17], Priestley and Taylor [5], Doorenbos and Pruitt [18], and Abtew [19]. The methods are described as follows:

2.2.1. Makkink method

$$ET = \alpha \times \left(\frac{\Delta}{\Delta+\gamma}\frac{R_s}{\lambda}\right) - \beta \tag{6}$$

R_s represents the amount of solar radiation (MJ m^{-2} d^{-1}); Δ is the slope of the saturated vapor pressure curve (kPa $°C^{-1}$); γ represents the humidity constant (kPa $°C^{-1}$); λ is the latent heat of evaporation (MJ kg^{-1}); and $\alpha = 0.61$, $\beta = 0.12$.

2.2.2. Turc method

1. Average relative humidity $RH < 50\%$

$$ET = 0.013\left(\frac{T}{T+15}\right) \times (R_s \times 23.8846 + 50) \times \left(1 + \frac{50 - RH}{70}\right) \quad (7)$$

2. Average relative humidity RH > 50%

$$ET = 0.013\left(\frac{T}{T+15}\right)(R_s \times 23.8846 + 50) \quad (8)$$

In Eq. (8), T represents the average temperature (°C); R_s is the amount of solar radiation (MJ m^{-2} d^{-1}); and RH represents average relative humidity (%).

2.2.3. *Jensen-Haise method*

$$ET = C_T \times (T - T_x) \times R_s \quad (9)$$

C_T represents the temperature constant, and its calculation method is listed below:

$$C_T = \frac{1}{(C_1 + C_2 \times C_H)} \quad (10)$$

$$C_1 = 68 - 3.6 \times \frac{h_j}{1000} \quad (11)$$

$$C_2 = 13 \quad (12)$$

$$C_h = \frac{50}{e_s(T_{max}) - e_s(T_{min})} \quad (13)$$

h_j is the sea surface height of the meteorological station; $e_s(T_{max}) - e_s(T_{min})$ represents the saturated vapor pressure at the highest temperature and the lowest temperature, respectively; T is the average temperature (°F); and T_x represents the temperature-axis intercept constant, and its formula is as follows:

$$T_x = 27.5 - 0.25 \times \left(e(T_{max}) - e(T_{min})\right) - \frac{h}{1000} \quad (14)$$

2.2.4. *Priestley-Taylor method*

$$ET = \alpha_{PT} \frac{\Delta}{\Delta + \gamma} \frac{R_n}{\lambda} \quad (15)$$

Δ represents the slope of the saturated vapor pressure curve (kPa °C^{-1}); γ is the humidity constant (kPa °C^{-1}); R_n is the net radiation (W m^{-2} d^{-1}); G represents soil thermal flux (MJ m^{-2} d^{-1}); and α_{PT} represents the empirical coefficient ($\alpha_{PT} = 1.26$).

2.2.5. Doorenbos-Pruitt method

$$ET = a + b \times \left(\frac{\Delta}{\Delta + \gamma} \frac{R_s}{\lambda}\right) \tag{16}$$

$$a = 1.066 - 0.13 \times 10^{-2} RH + 0.45\, U_z - 0.2 \times 10^{-3} RH \times U_z - 0.315 \times 10^{-4} RH^2 - 0.11 \times 10^{-2} U_z^2 \tag{17}$$

$$b = -0.3 \tag{18}$$

R_s is the amount of solar radiation (MJ m^{-2} d^{-1}); Δ represents the slope of the saturated vapor pressure curve (kPa $°C^{-1}$); γ is the humidity constant (kPa $°C^{-1}$); λ represents the latent heat of evaporation (MJ kg^{-1}); U_z is the wind speed (m s^{-1}); and RH represents relative humidity (%).

2.2.6. Abtew method

$$ET = \alpha \times \left(\frac{R_s}{\lambda}\right) \tag{19}$$

In Eq. (19), R_s represents the amount of solar radiation (MJ m^{-2} d^{-1}); λ represents the latent heat of evaporation (MJ kg^{-1}); and $\alpha = 0.53$.

2.3. Temperature-based methods

Temperature was the easiest to obtain among the many meteorological parameters. Generally speaking, the form of temperature-based methods is as follows [10]:

$$ET = c \times T^n \text{ or } ET = c \times d \times T(c_1 - c_2 h) \tag{20}$$

In Eq. (20), T is the air temperature (°C); h represents humidity; c, c_1, and c_2 were constants; and d represents time.

Four temperature-based methods were chosen in this study to estimate ET, including the Thornthwaite [20], Blaney and Criddle [21], Hamon [22], and Linacre [23]. The methods are described below:

2.3.1. Thornthwaite method

$$ET = C \times 16 \times \left(\frac{10T}{I}\right)^a \tag{21}$$

In Eq. (21), T represents monthly average temperature of the air (°C); I is the thermal index, and its formula is as follows:

$$I = \sum_{j=1}^{12} i_j \tag{22}$$

$$i = \left(\frac{T}{5}\right)^{1.51} \tag{23}$$

$$a = 0.000000675I^3 - 0.0000771I^2 + 0.0179I + 0.49239 \tag{24}$$

C represents the correction coefficient.

$$C = \frac{N}{360} \tag{25}$$

N represents monthly amount of daylight hours (h).

2.3.2. Blaney-Criddle method

$$ET = p \times (0.46T + 8.13) \tag{26}$$

P represents the annual daylight percentage of every month and T is the average temperature (°C).

2.3.3. Hamon method

$$ET = k \times 0.1651 \times 216.7 \times N \times \left(\frac{e_s}{T + 273.3}\right) \tag{27}$$

In Eq. (27), k represents the empirical coefficient (k = 1.0); N represents daylight hours (h); e_s is the saturated vapor pressure (kPa); and T represents average temperature (°C).

2.3.4. Linacre method

$$ET = \frac{\frac{500T_m}{100-A} + 15(T - T_d)}{(80 - T)} \tag{28}$$

$$T_m = T + 0.006h \tag{29}$$

T represents average temperature (°C); T_d is the dew point temperature (°C); and A represents latitude (°).

2.4. Statistical verification

In this study, the differences and correlations between the estimation results of the Penman-Monteith method and other formulas were compared and assessed using the following criteria:

2.4.1. Mean bias error

The bias degree of the Penman-Monteith method and the other methods was determined from the mean bias error (MBE). A smaller value indicated a lower bias degree as well as a better result. The best fit was MBE = 0, and the formula is as follows:

$$\mathrm{MBE} = \frac{\sum_{i=1}^{n} (E_i - P_i)}{n} \tag{30}$$

E_i represents the estimated value of the empirical formula; P_i represents the estimated value of the Penman-Monteith method; and n is the total number of observations.

2.4.2. Error percentage

$$\text{Error percentage} = \frac{\mathrm{MBE}}{\overline{x}} \times 100 \tag{31}$$

MBE represents the mean bias error of Eq. (30); and $\overline{x}$ represents the mean value.

2.4.3. Root mean square error

$$\mathrm{RMSE} = \sqrt{\frac{\sum_{i=1}^{n} (E_i - P_i)^2}{n}} \tag{32}$$

Root mean square error (RMSE) represents the variance degree of two estimated values. The best fit was RMSE = 0. In Eq. (32), E_i is the estimated value of empirical formula; P_i represents the estimated value of the Penman-Monteith method; and n is the total number of observations.

2.4.4. Pearson-type goodness-of-fit index (R^2)

$$R^2 = \left[\frac{\sum_{i=1}^{n} (E_i - \overline{E})(P_i - \overline{P})}{\sqrt{\sum_{i=1}^{n} (E_i - \overline{E})^2} \sqrt{\sum_{i=1}^{n} (P_i - \overline{P})^2}} \right] \tag{33}$$

The Pearson-type goodness-of-fit index represents the degree of correlation between two estimation methods. The best fit was R^2 = 1.0. In Eq. (33), E_i represents the estimated value of the empirical formula; $\overline{E}$ is the average estimated value of the empirical formulas; P_i represents the estimated value of the Penman-Monteith method; $\overline{P}$ is the mean estimated value of the Penman-Monteith method; and n represents the total number of observations.

2.5. Study area

There is abundant precipitation in Taiwan. Its distribution, however, is uneven in both time and space. In addition to the significant precipitation difference between the wet season and dry season, the high mountains and steep slopes in Taiwan have insufficient reservoir storage as well as ET losses that collectively result in an extremely low amount of usable water. Water resource management could be achieved by accurately estimating ET to predict available water resources. In this study, the meteorological data recorded during the period of 1961–2013 by the Tainan weather station of Taiwan and provided by the Central Weather Bureau were considered (**Figure 1**). The collected meteorological parameters included temperature, wind speed, relative humidity, solar radiation, vapor pressure difference, daylight hours, and so on. Because the climatic factors that influenced ET might change with variation in the time scale, previous researches suggested that average monthly data would lead to a better result [24]. Therefore, this study used average monthly data for estimation.

Figure 1. Location of Tainan weather station.

3. Results and discussion

3.1. Estimation of ET using Penman-Monteith method

The Penman-Monteith method is the main approach recommended internationally to estimate ET; it requires the use of meteorological parameters, such as radiation, air temperature, relative humidity, and wind speed. These parameters might be difficult to obtain and measure in many meteorological stations, with the exception of temperature. Using Taiwan as an example, only a few meteorological stations had complete data of all meteorological parameters, and still there were missing data in the observation materials. Yeh et al. [13] evaluated the ET difference between the Penman-Monteith method and evaporation pan in southern Taiwan. They used six meteorological stations in the southern part of Taiwan as case studies and collected meteorological data over a span of 15 years from 1990 to 2004 to estimate ET and ET of evaporation pans. In addition, a coefficient of evaporation pans was established. The results showed that the Penman-Monteith method and evaporation pan were highly correlated. Therefore, this study used long-term meteorological data from 1961 to 2013 from Tainan Weather Station provided by Central Weather Bureau for estimation. The estimation results calculated using the Penman-Monteith method were taken as the standard, which were named PM1. In cases where radiation data were missing or incomplete, Eq. (3) was used for substitution, which was called PM2. When wind speed data were missing or incomplete, the average wind speed of 1.83 m s^{-1} in Taiwan was used for calculation [15], which was called PM3. Eq. (2) was used for substitution in cases where relative humidity data were missing or incomplete, which was named PM4. Finally, when radiation, wind speed, and relative humidity data were all missing, all of the above substitutes were used, which was called PM5. Statistical methods of MBE, RMSE, and R^2 were applied to this study. In addition, the four models, namely PM2, PM3, PM4, and PM5, were used to estimate ET, and the results were compared with those of model PM1.

The characteristics of the ET at Tainan Weather Station estimated using different models are shown in **Table 1**. This demonstrates that maximum values are mainly concentrated in July, while minimum values are primarily concentrated in January or December. The average value was within the range of 3.42–3.61 mm/day. In addition, the ET estimated by PM models at Tainan Weather Station is shown in **Figure 2**, and the results indicate that the trend of each PM model was roughly the same as that of PM1. The comparison results between each PM model and PM1 are shown in **Figure 3**. This suggests that ET was underestimated by PM2 from July to September, while it was overestimated during other months; PM3 underestimated ET in

Scenarios	Min.	Min. (month)	Max.	Max. (month)	Mean	Standard deviation
PM1	2.26	January	4.72	July	3.54	0.92
PM2	2.35	January	4.55	July	3.61	0.85
PM3	2.22	January	4.78	July	3.58	0.95
PM4	2.15	December	4.56	July	3.42	0.89
PM5	2.24	December	4.38	July	3.49	0.81

Table 1. Various scenarios for calculating evapotranspiration using the Penman-Monteith method (mm d^{-1}).

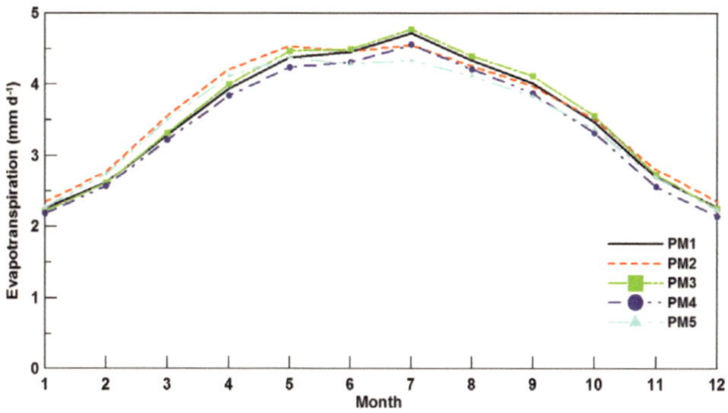

Figure 2. Comparison of monthly mean values of the Penman-Monteith method from various scenarios.

January, February, and December, and it was overestimated in the remaining months; ET was underestimated by PM4 in all months; PM5 underestimated ET in June, July, August, September, October, and December and overestimated ET in other months.

In this study, the estimated ET of PM2, PM3, PM4, and PM5 models were compared with that of the PM1 model using the statistical methods of MBE, RMSE, and R^2. Statistical verification

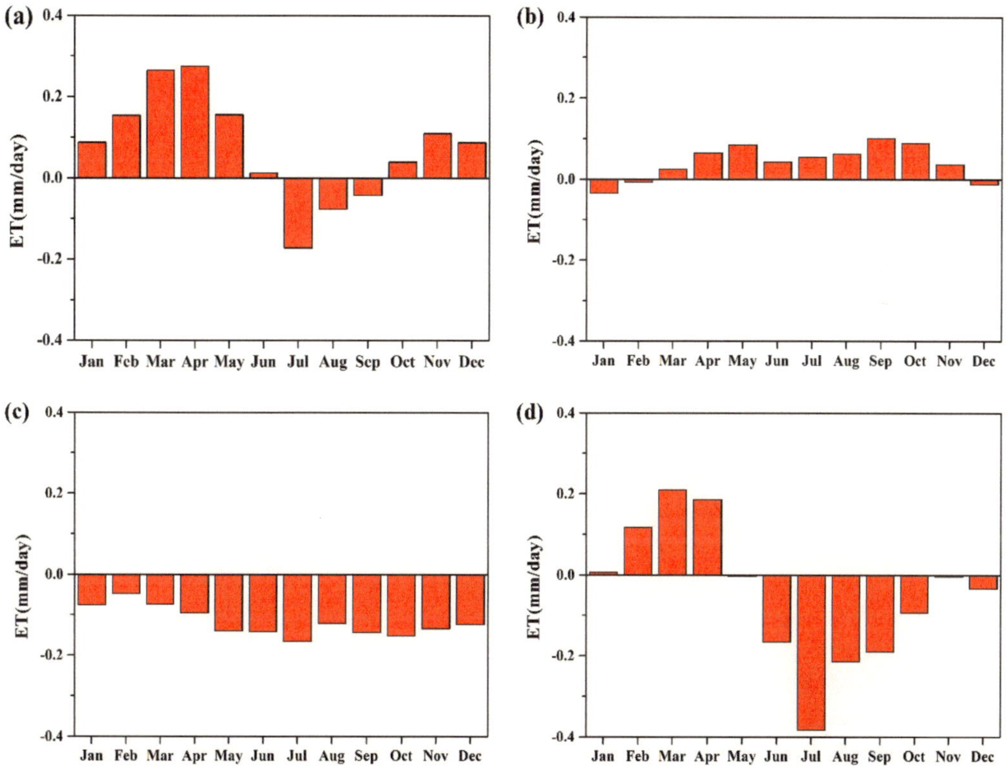

Figure 3. Monthly evapotranspiration comparison between the PM1 and the calculated values by using the various scenarios. (a) PM2; (b) PM3; (c) PM4; (d) PM5.

results of MBE showed that the MBE value was within the range of −0.12 to 0.07 mm d^{-1}, while the value of RMSE ranged from 0.09 to 0.31 mm d^{-1}. There is slight overestimation in PM2 and PM3. In contrast, there is slight underestimation in PM4 and PM5. In addition, R^2 was optimal in PM3 based on the statistical verification results. Therefore, according to the calculation results of MBE, RMSE, and R^2, the PM3 model had the optimal performance. Given the above comparison, the results of this study showed that wind speed had little effect on the assessment of ET with the Penman-Monteith method as the standard (PM1), which was similar to the conclusion drawn by Jabloun et al. [25]. The results obtained by Popova et al. [26] using the global average wind speed of 2 m s^{-1} were similar to PM1 as well. In addition, compared to missing radiation or wind speed data, the absence of relative humidity data exerted a larger impact on the estimation of ET when Penman-Monteith method was used.

3.2. The results of ET estimation using different empirical formulas

Compared with other meteorological parameters, namely radiation, relative humidity, and wind speed, temperature is relatively easy to obtain. Apart from that, radiation can be accurately measured, and yet existing measurement methods are unable to acquire precise wind speed data, especially in dry areas where the error would be relatively larger. Because the mixed evaluation methods, such as the Penman-Monteith method, require many meteorological parameters, there are some difficulties in the funding, maintenance, and construction of meteorological stations, making it difficult to acquire certain data. Therefore, it is essential to develop ET estimation methods that require fewer or a single meteorological parameter [27]. A number of scholars have proposed various methods or experiential formulas and compared them to the Penman-Monteith method in the hope of finding a relatively simple method and experiential formula to measure ET [28]. This study selected six radiation-based methods and four temperature-based methods to explore their applicability in the study area.

3.2.1. Estimation of monthly average ET using radiation-based methods

According to the radiation-based estimation methods that are used internationally, this study selected six methods, including Makkink [4], Turc [16], Jensen and Haise [17], Priestley and Taylor [5], Doorenbos and Pruitt [18], and Abtew [19]. A commonly used statistical mean error percentage was applied to the estimation so as to discuss the basic statistical differences. The data recorded by Tainan Weather Station from 1961 to 2013 were substituted into the formula, and the results are shown in **Table 2**. This demonstrates that minimum values were mainly concentrated in December and January, while maximum values were primarily concentrated in July. The mean value indicated a significant underestimation in ET calculated by the Makkink method, with an average value of 2.99 mm d^{-1} and an error percentage of −15.5%. The results of the Turc method showed a slight overestimation, with an average value of 3.66 mm d^{-1} and an error percentage of 3.4%. ET was significantly overestimated by the Jensen-Haise method, with a mean value of 5.16 mm d^{-1} and an error percentage of 45.8%. The results of the Priestley-Taylor method suggested an overestimation, with an average value of 3.96 mm d^{-1} and an error percentage of 11.9%. ET calculated by the Doorenbos-Pruitt method was the closest to the mean value of the Penman-Monteith method, with an average value of 3.43 mm d^{-1} and an

	Min.	Min. (month)	Max.	Max. (month)	Mean	Standard deviation
Penman-Monteith	2.26	January	4.72	July	3.54	0.92
Makkink (1957)	1.95	December	3.92	July	2.99	0.74
Turc (1961)	2.51	December	4.67	July	3.66	0.82
Jensen-Haise (1963)	2.91	January	7.28	July	5.16	1.63
Priestley-Taylor (1972)	2.16	December	5.53	July	3.96	1.23
Doorenbos-Pruitt (1977)	2.64	December	5.27	July	3.43	3.68
Abtew (1996)	2.69	December	4.50	July	3.68	0.70

Table 2. Performance evaluation of the radiation-based methods against Penman-Monteith (mm d^{-1}).

error percentage of −3.1%. The results of the Abtew method showed a slight overestimation, with an average value of 3.68 mm d^{-1} and an error percentage of 4.0%.

The trend of monthly average ET at Tainan Weather Station calculated by various radiation-based methods were all consistent with that of the Penman-Monteith method, which was taken as the standard, as shown in **Figure 4**. As shown in **Figure 5**, the monthly average ET was underestimated by the Makkink method, with an error percentage ranging from −16.9 to −13.7%; the Turc method slightly overestimated all the monthly average ET, with an error percentage of −1.1 to 11.1%; and the monthly average ET was significantly overestimated by the Jensen-Haise method. Especially in summer, the overestimation was far more significant, and the error percentage was up to 54.2%. The results of the Priestley-Taylor method suggest underestimation only in December and January, and overestimation in other months, with an error percentage of −4.4 to 17.2%. The Doorenbos-Pruitt method slightly underestimated ET in May, August, and September, while in other months ET was overestimated with an error percentage ranging from 11.7 to 16.8%. Compared with the Penman-Monteith method, ET was overestimated by the Abtew method, and the error percentage was within the range of −4.7 to 19%. The above results suggest that the Doorenbos-Pruitt method was the least biased in estimating ET, while the Jensen-Haise method was the most biased.

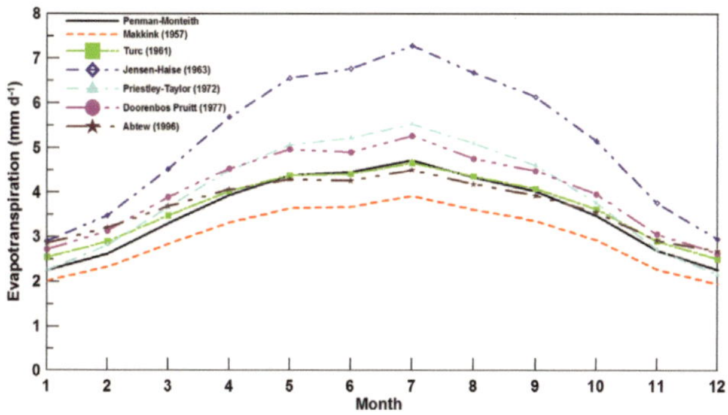

Figure 4. Monthly evapotranspiration computed by the Penman-Monteith method and six radiation-based methods.

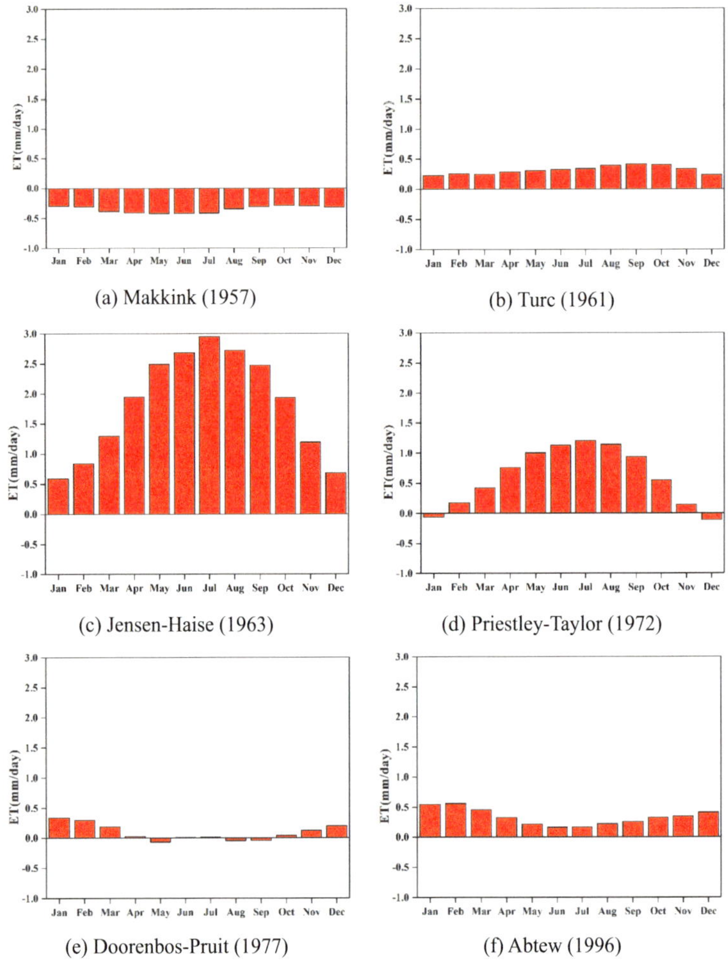

Figure 5. Monthly evapotranspiration comparison between the Penman-Monteith method and the calculated values by using the radiation-based methods ((a) Makkink; (b) Turc; (c) Jensen-Haise; (d) Priestley-Taylor; (e) Doorenbos-Pruit; (f) Abtew).

In addition, three statistical methods, namely MBE, RMSE, and R^2, were used in this study to compare the estimation results of the Makkink, Turc, Jensen-Haise, Priestley-Taylor, Doorenbos-Pruitt, and Abtew methods with the Penman-Monteith method. The statistical verification results of MBE indicated that the value of MBE was within the range of −0.55 to 0.41 mm d^{-1} and the value of RMSE ranged from 0.23 to 1.78 mm d^{-1}. **Figure 6(a)** shows that the Makkink method underestimated ET, and the MBE value was −0.55. Furthermore, the other five methods, Turc, Jensen-Haise, Priestley-Taylor, Doorenbos-Pruitt, and Abtew methods, all overestimated ET, and the MBE values were respectively 0.12, 1.62, 0.41, 0.49, and 0.14. In particular, the overestimation of the Jensen-Haise method was the most significant. **Figure 6(b)** suggests that all six methods overestimated ET, and the values of RMSE were respectively 0.60, 0.23, 1.78, 0.55,

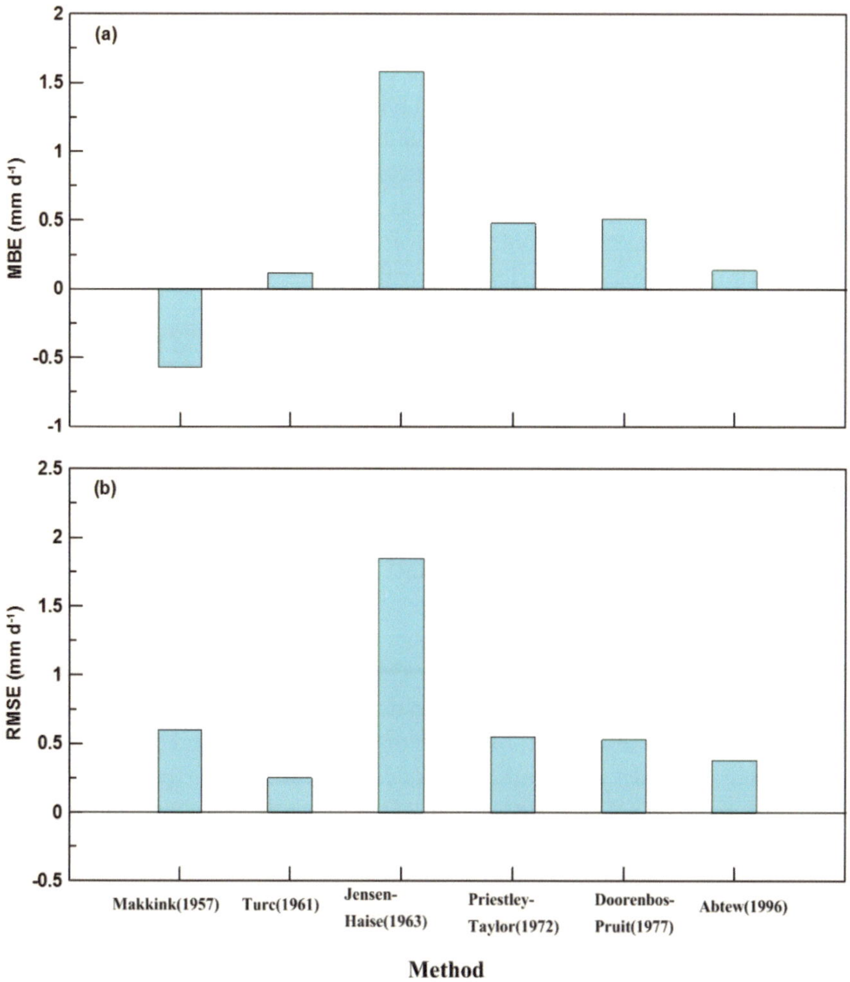

Figure 6. (a) MBE and (b) RMSE for evapotranspiration comparison between the Penman-Monteith method and six radiation-based methods.

0.53, and 0.37. Evapotranspiration was most significantly overestimated by the Jensen-Haise method as well. The statistical results showed that R^2 was within the range of 0.90–0.97. Therefore, judging from the statistical results of MBE, RMSE, and R^2, the Turc method was optimal at the Tainan Weather Station, followed by the Abtew method; the method with the worst performance was the Jensen-Haise method. Previously, Tabari et al. [29] used 31 methods to evaluate ET at a meteorological station named Rasht in a humid area of Iran. The results showed that, compared to the Penman-Monteith method, the Jensen-Haise method severely overestimated ET with a relative error of about 30%. It was also found to significantly overestimate ET in this study area, and the relative errors were respectively around 59 and 48%. Such overestimation also occurred in the humid regions of Serbia [30] and Florida of the USA [31].

3.2.2. Estimation of monthly average ET using temperature-based methods

Among the temperature-based methods that are commonly used internationally, this chapter selected four methods, including Thornthwaite [20], Blaney and Criddle [21], Hamon [22], and Linacre [23]. To begin with, the commonly used statistical mean value and error percentage were used for estimation; then this chapter discusses the basic statistical error. After the data recorded by Tainan Weather Station from 1961 to 2013 were substituted into the formulas to calculate the daily ET, the average monthly ET was calculated with month as the unit, and the results are shown in **Table 3**. At the Tainan Weather Station, minimum values were mainly concentrated in January and maximum values were primarily concentrated in July. The mean value suggests that the Thornthwaite method severely underestimated ET, with a mean value of 1.95 mm d^{-1} and an error percentage of −44.9%. The Blaney-Criddle method significantly underestimated ET as well. Its mean value was 1.61 mm d^{-1} and the error percentage was −54.5%. The Hamon method underestimated ET, with a mean value of 2.72 mm d^{-1} and an error percentage of −23.2%. Evapotranspiration was overestimated by the Linacre method, with a mean value of 4.05 mm d^{-1} and an error percentage of 14.4%.

This chapter compared the ET of Tainan Weather Station as calculated by the temperature-based methods with that of the Penman-Monteith method, and the results are as shown in **Figure 7**. The trends of the Thornthwaite, Hamon, and Linacre methods were consistent with the Penman-Monteith method, while Blaney-Criddle method suggested otherwise. The monthly average ET at Tainan Weather Station estimated by the temperature-based estimation methods was compared with Penman-Monteith method, as shown in **Figure 8**. Evapotranspiration was underestimated by Thornthwaite method, with an error percentage of −70.4 to 31.1%. The maximum error occurred in March and April. The Blaney-Criddle method underestimated monthly average ET, and the error percentage was within the range of −60.6 to 38.9%. Evapotranspiration was also underestimated by the Hamon method. The underestimation in winter was insignificant, with an error percentage of −23.5 to 18.1%. In May, the percentage reached its maximum. The Linacre method overestimated the monthly average ET. The error percentage ranged from 5.1 to 23.9% and reached maximum value in November. In light of the above results, the error percentage of Thornthwaite method was the largest.

This study used three statistical methods, namely MBE, RMSE, and R^2, to compare ET at the Tainan Weather Station estimated by the Thornthwaite, Blaney-Criddle, Hamon methods with that of the Penman-Monteith method. The value of MBE ranged from −1.93 to 0.51 mm d^{-1} and RMSE was within the range of 0.63–2.08 mm d^{-1}. The statistical results of R^2 indicated that

	Min.	Min. (month)	Max.	Max. (month)	Mean	Standard deviation
Penman-Monteith	2.26	January	4.72	July	3.54	0.92
Thornthwaite (1948)	0.67	January	3.25	July	1.95	1.02
Blaney-Criddle (1959)	1.38	January	1.86	July	1.61	0.32
Hamon (1961)	1.85	January	3.61	July	2.72	0.80
Linacre (1977)	2.80	January	4.96	July	4.05	0.82

Table 3. Performance evaluation of the temperature-based methods against Penman-Monteith (mm d^{-1}).

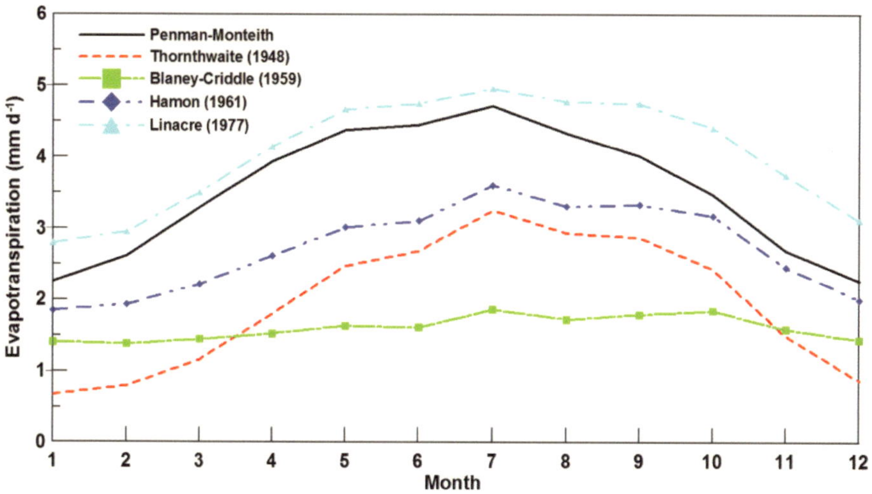

Figure 7. Monthly evapotranspiration computed by the Penman-Monteith method and four temperature-based methods.

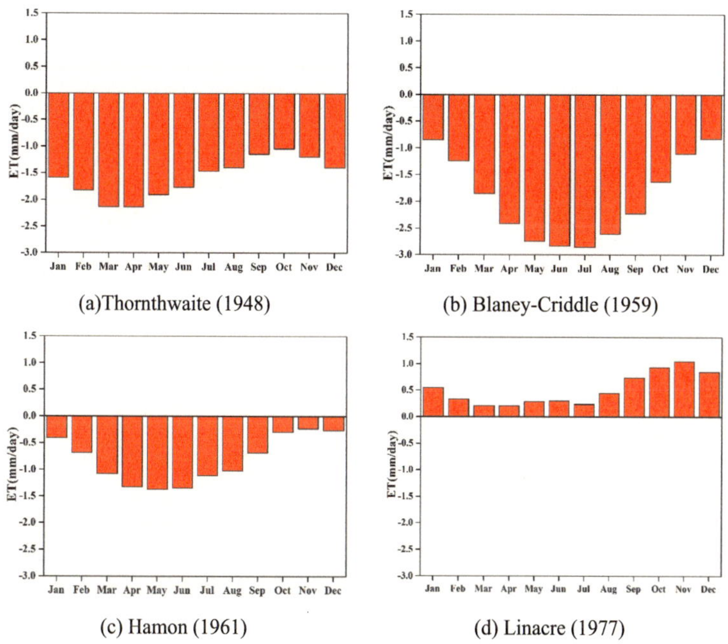

Figure 8. Monthly evapotranspiration comparison between the Penman-Monteith method and the calculated values by using the temperature-based methods ((a) Thornthwaite; (b) Blaney-Criddle; (c) Hamon; (d) Linacre).

it was within the range of 0.36–0.83. As shown in **Figure 9(a)**, the results of Thornthwaite, Blaney-Criddle, and Hamon methods suggest underestimation, and the values of MBE were −1.58, −1.93, and −0.82, respectively. Results of the Linacre method, however, indicated overestimation with an MBE of 0.51. **Figure 9(b)** suggests overestimation in Thornthwaite,

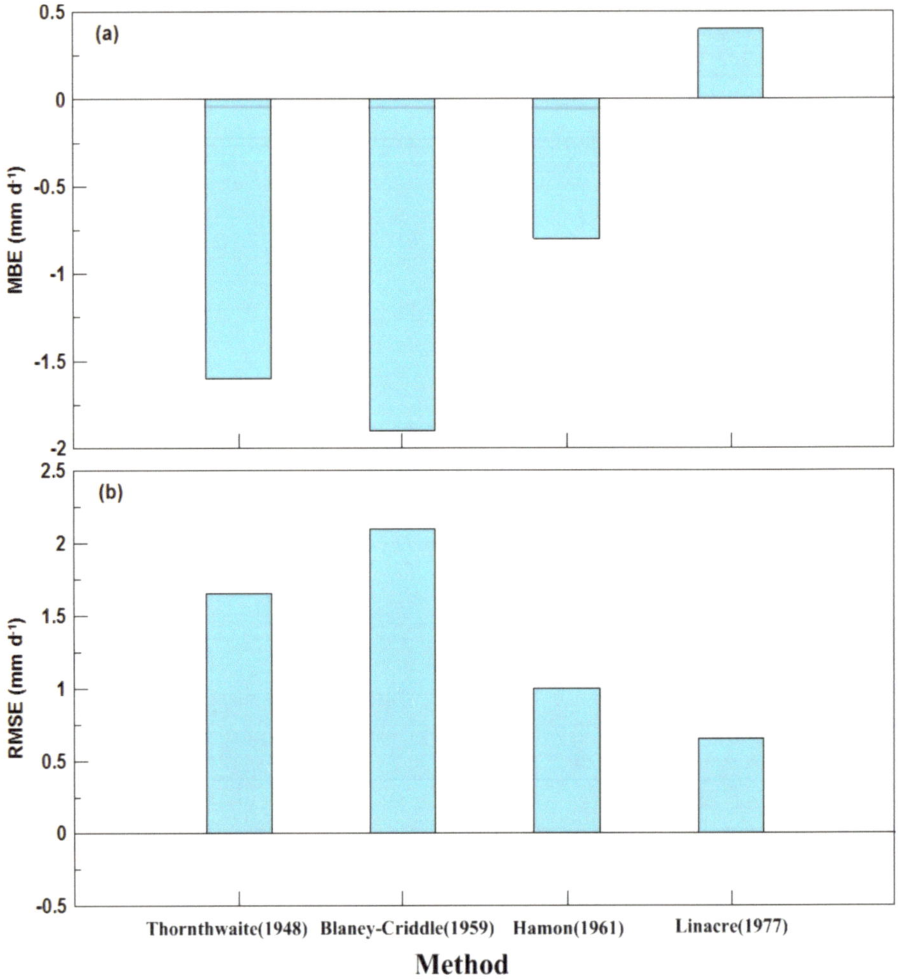

Figure 9. (a) MBE and (b) RMSE for evapotranspiration comparison between the Penman-Monteith method and four temperature-based methods.

Blaney-Criddle, Hamon, and Linacre methods, with RMSE values of 1.63, 1.31, 1.15, and 1.12, respectively. In summary, according to the statistical results of MBE, RMSE, and R^2, the Linacre method was optimal for estimating ET at the Tainan Weather Station, followed by the Hamon method. The Blaney-Criddle method was the least fit.

According to relevant studies and literature, Fontenot [32] declared that for meteorological stations near the coast, the Linacre method overestimated ET by 18.46% compared to the Penman-Monteith method. It was also pointed out that this method could be greatly affected by the dew point temperature. Compared with the Penman-Monteith method, the results of Thornthwaite, Hamon, and Blaney-Criddle methods all suggest underestimation, as these three temperature-based formulas all took daylight hours into consideration. In spite of the high temperature, the results would still be lower than the actual amount when daylight hours were insufficient, causing underestimation. Even if the daylight hours were insufficient, ET still occurred. The results of this study show that the Blaney-Criddle method underestimated ET in Tainan because it is strongly influenced by the annual daylight percentage of every month. Cruff and Thompson [33] used the Thornthwaite and Blaney-Criddle methods to estimate ET in the desert areas of the southwestern United States, and the results suggested underestimation as well.

This study compared the results of radiation-based methods with that of Penman-Monteith method and discovered that the empirical formulas of radiation-based methods were better than those of temperature-based methods. In addition, the errors of ET calculated by temperature-based methods were larger than those of the radiation-based methods. The reason is as follows: it is most likely that temperature is the only meteorological parameter used in empirical formulas of temperature-based methods. Therefore, it could be easily affected by the data of meteorological station, which would easily cause inaccuracy. Such a conclusion is similar to that of Lu et al. [34], Sentelhas et al. [8], and Gebhart et al. [35]. Moreover, the estimation results of Tukimat et al. [36] in Malaysia showed that three radiation-based methods, namely Makkink [4], Turc [16], and Priestley and Taylor [5], were more accurate than two temperature-based methods, the Thornthwaite [20] and Blaney and Criddle [21] methods. In terms of temperature-based estimation methods, many scholars have found that ET was underestimated by the Thornthwaite [20] method in humid areas compared to the Penman-Monteith method. For instance, the results of Alkaeed et al. [37] in Fukuoka of Japan, Trajkovic and Kolakovic [30] in six meteorological stations of Balkan Peninsula, and Sentelhas et al. [8] in Ontario of Canada, all showed the same conclusion. Some scholars, however, have pointed out that compared with the Penman-Monteith method, the performance of R^2 in the Thornthwaite [20] method was worse, and yet its trend was consistent with the Penman-Monteith method. The evaluation of ET conducted by [36] in Kedah of Malaysia suggested same result.

4. Conclusions

This study mainly aimed to estimate ET using a limited number of meteorological parameters. With the internationally accepted Penman-Monteith method as the standard, the estimation formulas of six radiation-based methods were compared with those of four temperature-based methods. The 53-year dataset recorded by Tainan Weather Station from 1961 to 2013 was used to discuss ET. Statistical indexes were used to analyze and discuss the differences in ET calculated by the Penman-Monteith method and other estimation formulas in the hope of discovering a simple estimation formula to solve the issue of lacking or missing meteorological data.

This study discussed situations in which meteorological data were insufficient or missing in the Penman-Monteith method. The results showed that using the average Taiwan wind speed of 1.83 m s^{-1} when wind speed data were insufficient or missing exerted little impact on ET estimation of the Penman-Monteith method. In the cases where empirical formulas were used for substitution because of the lack of relative humidity data, the estimated ET was higher than the actual data, causing overestimation. In addition, this study explored the impact on ET estimation by the Penman-Monteith method caused by insufficient or missing radiation, relative humidity, or wind speed data. It was discovered that the impact of wind speed was minimal, and the impact of relative humidity was the highest.

The six radiation-based methods selected in this study all suggested overestimation. In particular, the Turc method was optimal, followed by the Doorenbos-Pruitt method; the method with the worst performance was Jensen-Haise. This study found that ET was overestimated by the Jensen-Haise method in humid areas. In addition, among the four chosen temperature-based methods in this study, the Thornthwaite method, Hamon method, and Blaney-Criddle method all underestimated ET compared with the Penman-Monteith method, as these three temperature-based formulas all take daylight hours into consideration. In the cases where the daylight hours were insufficient, no matter how high the temperature was, underestimation would still occur. Even though the daylight hours were insufficient, ET was still occurring. The performance of the Linacre method was the best among the four estimation methods. The results of this study indicate that radiation-based estimation methods are better than temperature-based methods, as temperature is most likely to be the only meteorological parameter required in empirical formulas of temperature-based methods, making it easily affected by the data of meteorological stations, thus resulting in inaccuracy.

Author details

Hsin-Fu Yeh

Address all correspondence to: hfyeh@mail.ncku.edu.tw

Department of Resources Engineering, National Cheng Kung University, Tainan, Taiwan

References

[1] Sabziparvar AA, Mirmasoudi SH, Tabari H, Nazemosadat MJ, Maryanaji Z. ENSO teleconnection impacts on reference evapotranspiration variability in some warm climates of Iran. International Journal of Climatology. 2011;**31**:1710-1723. DOI: 10.1002/joc.2187

[2] Sumner DM, Jacobs JM. Utility of Penman-Monteith, Priestley-Taylor, reference evapotranspiration, and Pan evaporation methods to estimate pasture evapotranspiration. Journal of Hydrology. 2005;**308**:81-104. DOI: 10.1016/j.jhydrol.2004.10.023

[3] Temesgen B, Eching S, Davidoff B, Frame K. Comparison of some reference evapotranspiration equations for California. Journal of Irrigation and Drainage Engineering. 2005;**131**:73-84. DOI: 10.1061/(ASCE)0733-9437(2005)131:1(73)

[4] Makkink GF. Testing the Penman formula by means of lysimeters. Journal of the Institution of Water Engineering. 1957;**11**:277-288

[5] Priestley C, Taylor R. On the assessment of surface heat flux and evaporation using large-scale parameters. Monthly Weather Review. 1972;**100**:81-92. DOI: 10.1175/1520-0493(1972)100<0081:OTAOSH>2.3.CO;2

[6] Gavilan P, Berengena J, Allen RG. Measuring versus estimating net radiation and soil heat flux: Impact on Penman–Monteith reference ET estimates in semiarid regions. Agricultural Water Management. 2006;**89**:275-286. DOI: 10.1016/j.agwat.2007.01.014

[7] Drexler JZ, Snyder RL, Spano D, Paw KTU. A review of models and micrometeorological methods used to estimate wetland evapotranspiration. Hydrology Processes. 2004;**18**:2071-2101. DOI: 10.1002/hyp.1462

[8] Sentelhas PC, Gillespie TJ, Santos EA. Evaluation of FAO Penman-Monteith and alternative methods for estimating reference evapotranspiration with missing data in Southern Ontario, Canada. Agricultural Water Management. 2010;**97**:635-644. DOI: 10.1016/j.agwat.2009.12.001

[9] Tabari H, Talaee PH. Local calibration of the Hargreaves and Priestley-Taylor equations for estimating reference evapotranspiration in arid and cold climates of Iran based on the Penman-Monteith model. Journal of Hydrologic Engineering. 2011;**16**:837-845. DOI: 10.1061/(ASCE)HE.1943-5584.0000366

[10] Xu CY, Singh VP. Cross comparison of empirical equations for calculating potential evapotranspiration with data from Switzerland. Water Resources Management. 2002;**16**:197-219. DOI: 10.1023/A:1020282515975

[11] Cristea NC, Kampf SK, Burges SJ. Revised coefficients for Priestley-Taylor and Makkink-Hansen equations for estimating daily reference evapotranspiration. Journal of Hydrologic Engineering. 2013;**18**:1289-1300. DOI: 10.1061/(ASCE)HE.1943-5584.0000679

[12] Allen RG, Periera LS, Raes D, Smith M. Crop Evapotranspiration: Guideline for Computing Crop\Water Requirement, FAO Irrigation and drainage paper 56. FAO: Rome, 300, 1998. p. 6541

[13] Yeh HF, Lee CH, Chen CW, Chang KL. Evaluation of pan coefficients for estimating reference evapotranspiration in southern Taiwan. Journal of Taiwan Agricultural Engineering. 2008;**54**:27-35 (in Chinese with English abstract)

[14] Yeh HF, Lee CH, Chen JF. Optimum comparison of empirical equations for calculating potential evapotranspiration. Journal of Taiwan Agricultural Engineering. 2005;**51**:27-37 (in Chinese with English abstract)

[15] Yeh HF, Lin HI, Lee CH. Assessment of spatial and temporal distribution of reference evapotranspiration and pan evaporation in Taiwan. Journal of Taiwan Agricultural Engineering. 2013;**59**:13-21 (in Chinese with English abstract)

[16] Turc L. Evaluation des besoins en eau d'irrigation, evapotranspiration potentielle, formule climatique simplifee et mise a jour. Annal Agron. 1961;**12**:13-49

[17] Jensen ME, Haise HR. Estimation of evapotranspiration from solar radiation. Journal of Irrigation and Drainage Division. 1963;**89**:15-41

[18] Doorenbos J, Pruitt WO. Guidelines for Predicting Crop Water Requirements. FAO Irrigation and Drainage Paper 24, Food and Agriculture Organization: Rome. 1977

[19] Abtew W. Evapotranspiration measurements and modeling for three wetland systems in south Florida. Water Resources Bulletin. 1996;**32**:465-473. DOI: 10.1111/j.1752-1688.1996.tb04044.x

[20] Thornthwaite CW. An approach toward a rational classification of climate. Geographical Review. 1948;**38**:55-94

[21] Blaney HF, Criddle WD. Determining water requirements in irrigated areas from climatological and irrigation data. Soil Conservation Service Technical Paper 96. Soil Conservation Service, U.S. Dept. of Agriculture: Washington, D.C. 1959

[22] Hamon WR. Estimating potential evapotranspiration. Journal of Hydraulics Division, Proceedings of the American Society of Civil Engineers. 1961;**87**:107-120

[23] Linacre ET. A simple formula for estimating evaporation rates in various climates, using temperature data alone. Agricultural Meteorology. 1977;**18**:409-424. DOI: 10.1016/0002-1571(77)90007-3

[24] Chang KL. Application of Pan Coefficients on Estimating Regional Reference Evapotranspiration. Master Thesis, Department of Resources Engineering, National Cheng Kung University: Tainan, Taiwan. 2007 (in Chinese with English abstract)

[25] Jabloun M, Sahli Erratum A. Evaluation of FAO-56 methodology for estimating reference evapotranspiration using limited climatic data: Application to Tunisia. Agricultural Water Management. 2008;**98**:707-715. DOI: 10.1016/j.agwat.2008.01.009

[26] Popova Z, Kercheva M, Pereira LS. Validation of the FAO methodology for computing ET with limited data application to south Bulgaria. Journal of Irrigation and Drainage Engineering. 2006;**55**:201-215. DOI: 10.1016/j.agwat.2008.01.009

[27] Pereira AR, Pruitt WO. Adaptation of the Thornthwaite scheme for estimating daily reference evapotranspiration. Agricultural Water Management. 2004;**66**:251-257. DOI: 10.1016/j.agwat.2003.11.003

[28] Fooladmand HR, Zandilak H, Ravanan MH. Comparison of different types of Hargreaves equation for estimating monthly evapotranspiration in the south of Iran. Archives of Agronomy and Soil Science. 2008;**54**:321-330. DOI: 10.1080/03650340701793603

[29] Tabari H, Grismer ME, Trajkovic S. Comparative analysis of 31 reference evapotranspiration methods under humid conditions. Irrigation Science. 2013;**31**:107-117. DOI: 10.1007/s00271-011-0295-z

[30] Trajkovic S, Kolakovic S. Evaluation of reference evapotranspiration equations under humid conditions. Water Resources Management. 2009;**23**:3057-3067. DOI: 10.1007/s11269-009-9423-4

[31] Irmak S, Allen RG, Whitty EB. Daily grass and alfalfa reference evapotranspiration estimates and alfalfa-to-grass evapotranspiration ratios in Florida. Journal of Irrigation and Drainage Engineering. 2003;**129**:360-370. DOI: 10.1061/(ASCE)0733-9437(2003)129:5(360)

[32] Fontenot RL. An Evaluation of Reference Evapotranspiration Models in Louisiana [master thesis]. Louisiana State University: USA. 2004.

[33] Cruff RW, Thompson TH. Comparison of Methods of Estimating Potential Evapotranspiration from Climatological Data in Arid and Subhumid Environments. USGS water-supply paper: USA, 1839, 1967.

[34] Lu JB, Sun G, McNulty SG, Amatya DM. A comparison of six potential evapotranspiration methods for regional use in the southeastern United States. Journal of the American Water Resources Association. 2005;**41**:621-633. DOI: 10.1111/j.1752-1688.2005.tb03759.x

[35] Gebhart S, Radoglou K, Chalivopoulos G, Matzarakis A. Evaluation of potential evapotranspiration in central Macedonia by EmPEst. In: Helmis CG, Nastos PT, editors. Advances in Meteorology, Climatology and Atmospheric Physics. Springer. 2013. pp. 451-456

[36] Tukimat NNA, Harun S, Shahid S. Comparison of different methods in estimating potential evapotranspiration at Muda Irrigation Scheme of Malaysia. Journal of Agriculture and Rural Development in the Tropics and Subtropics. 2012;**113**:77-85

[37] Alkaeed O, Flores C, Jinno K, Tsutsumi A. Comparison of several reference evapotranspiration methods for Itoshima Peninsula area, Fukuoka, Japan. Memoirs of the Faculty of Engineering, Kyushu University. 2006;**66**:1-14

Chapter 2

Assessment and Prediction of Evapotranspiration Based on Scintillometry and Meteorological Datasets

Antonin Poisson, Angel Fernandez, Dario G. Gomez, Régis Barillé and Benoit Chorro

Additional information is available at the end of the chapter

http://dx.doi.org/10.5772/intechopen.68538

Abstract

Two methods are used for estimating the evapotranspiration (ET) rate: scintillometry and meteorological measurements using the FAO-PM56 model with the reference evapotranspiration for the crop (ETO) and the specific coefficient (Kc) for corn at its stage development. Measurements were done on a field with homogeneous corn crop at the stage of 3 months before the final harvest (65 % of maximum plant growth). The two methods are compared with environmental parameters to determine the most influential on the final result of ET.

A coefficient of 0.78 is found between the two methods resulting of an underestimation of the evapotranspiration values with FAO.

The sensitivity for the two measurements are compared in order to determine how sensitive the output calculation of evapotranspiration could be with respect to the calculation elements which are subject to uncertainty of variability in the input environmental parameters. The scintillometer uncertainty is lower than the FAO-56 uncertainty.

Finally, a model based on an artificial neural network (ANN) forecasting ET is developed in order to anticipate the necessary action for water management. It leads to the conclusion that scintillometry is more able to predict evapotranspiration on short and medium time than the FAO-PM56 method.

Keywords: evapotranspiration, FAO-PM56, scintillometry, artificial neural network

1. Introduction

Water resources are highly influenced by the hydrologic cycle and play a role in the agriculture economic development. However, as it is shown by the intergovernmental panel on climate change report [1], the phenomenon of changing climate is set way to exacerbate an already serious situation of water supply for various users. Agricultural production will be one of the sectors most vulnerable to climate change and variability. The water budget must now be shared with agriculture, urban use, industry, recreation and livestock watering; the future will be seeing an increasing competition for water. Spatial and temporal changes in precipitation and temperature patterns will have an impact on the viability of dry land farming and therefore necessitate irrigation where rainfall was previously adequate. Efficiency improvement in irrigation lies among the key strategies for saving more water and promote a sustainable intensification of agriculture when water scarcity becomes a major constraint to production [2]. Nevertheless, irrigation water for crops is globally the major consumptive use of water resources. Due to the above-mentioned challenges, it is important to improve the management of agricultural water, which would involve the accurate estimation of consumptive uses. One of the techniques is the measurement of evapotranspiration which is a major component of the hydrologic cycle. Evapotranspiration (ET) is an essential component of the water balance, and it is a significant consumptive witness of precipitation and water applied for irrigation of cropland [3]. ET can help for highly efficient management of water uses in agriculture and set up real water-saving systems.

Basically, ET includes two processes: One is evaporation and the second is transpiration. The latter is the process of removing water from vegetation or any other moisture containing living surface. Evapotranspiration includes two processes. During the plant growth, the water stored in the soil is taped and transferred in the atmosphere. Transpiration is the evaporation of water in the vascular system of plants through the leaf stomata when they open and close controlled by their guard cells. Based on this bio-physical process, transpiration involves a living organism and its tissues. ET is then the process, whereby water originating from a wide range of sources is transferred from the soil compartment and/or vegetation layer to the atmosphere. ET is the largest outgoing water flux from the Earth's surface and accurately quantifying ET is critical for the development of crop cultures in an increasing drier environment, and it can contribute to a greater understanding of a range of agricultural ecosystem processes. ET is particularly fundamental when dealing with water resource management issues such as irrigation water or water reserve management [4]. ET cannot be directly measured but it has to be estimated by monitoring the exchange of energy/water above the vegetated surface (remote sensing) or as a residual term of the hydrological balance.

Several methods are currently used to measure and estimate ET: One of them is the lysimeter method or soil water budget. That method may be accurate but lysimeters are expensive, and the extent of their measurement is localized (i.e., they provide data for a very small area compared to the field surface, so it can only be used in field locations). Another one uses micrometeorological data to compute ET. A widely used approach by these data is the FAO-24 and by extension the FAO-56 procedure, based on ET_0 and Kc [5]. ET acquisition can be obtained with different instruments at the scale of: the leaf (porometer), an individual plant (i.e., sapflow, lysimeter), the field scale (i.e., field water balance, Bowen ratio, scintillometer) and the landscape scale (i.e., eddy correlation and catchment water balance) [6]. The flux measurement of micro-meteorological station can only represent the value in a point or a limited area (several meters to several hundred meters). However, a scintillometer can measure averaged sensible heat fluxes in a distance of 500m to 10 km, which is an average of time and space. The measurement scale of a scintillometer is matching to the grid scale of atmospheric model and the pixel scale of remote sensing, as a result. This advantage promotes the development of scintillometers in recent years [7].

The objective of this work is to compare two methods used for estimating ET rate: scintillometry and meteorological measurements with the FAO-PM56 model based on ET_0 and Kc with the reference evapotranspiration for the crop (ET_0) and the specific crop coefficient for the cultivation type at its stage development (Kc). Also, it is compared how the final result of ET, calculated with the two different measurements, can be more sensible to different environmental parameters. The sensitivity of the two methods is calculated and the influence of the main environmental parameters on the accurate values of ET.

In order to anticipate the necessary action for water management, a model to forecast ET based on an artificial neural network (ANN) is developed. In recent years, ANN models have become extremely popular for prediction and forecasting in a number [8, 9] of domains, including finance, power generation, medicine, water resources and environmental science [8, 9]. The evapotranspiration process calculated with the FAO-PM56 and scintillometry data is a nonlinear process. ANN models are quite appropriate for the simulation of ET leading to good results.

The predicted output final results are compared for two different input parameters of ET calculated with scintillometry and micro-meteorology data. The final goal of this study is to find which input data can be reliable to obtain ET forecast performances. Moreover, the optimal number of predicted days to obtain a correct final performance and the optimal number of input days of data to obtain a correct prediction are tested.

2. Materials and methods

2.1. Site description

The study was carried out in the west part of France near to Niort (France). The site of the intercomparison of evapotranspiration measurements is located in the village of Sainte Soline (contained near 46°15′27.7″N—0°02′32.0″E) (**Figure 1**), and the area's elevation is 117—145 m above the sea level. According to data from the Meteorologisk Institute, the daily mean temperatures vary significantly from 5°C in January to 26°C in July. The average days with precipitation per month are 12 days in January and 6 days in August. Mean annual rainfall is 6.4 mm and is distributed relatively evenly across all months. The dominant wind directions are from the south-west and west-south-west and north. The site consists of a field with homogeneous corn crop. Corn corresponds to plants DKc 4590. Corn was planted in April, and the measurements were made at the stage of 4 months (almost 5 months) after planting corresponding to about 65% of maximum plant growth and to 3 months (just before) of the final harvest.

The soil type is clay, and the typical porosity of this soil type is about 0.30 $m^3\ m^{-3}$ with grain size <0.002 mm.

Figure 1. Location map of the measurements during the summer time and contained between 46°15′27.7″N and 0°02′32.0″E.

2.2. Measurement description

The scintillometer provides the opportunity to obtain surface fluxes of sensible heat across a distance of several kilometers and over a heterogeneous landscape [10]. As shown by different authors, it is feasible to use the scintillometer for estimating area-averaged sensible heat flux <λE> ($\lambda E = R_n - G - H$) as the residual term of the energy balance equation, providing estimations of area-average available energy ($\lambda E = R_n - G$) with E the sensible heat flux, λ the latent heat, R_n the net radiation and G the soil heat flux [11]. The transmitter and receiver of the scintillometer system were installed at opposite edges of the field, and the electromagnetic radiation was transmitted across the field. Scintillometry measurements are based on the propagation of electromagnetic waves in atmosphere and the measurement of its disturbance by atmospheric turbulence. The turbulence effect induces laser beam fluctuations leading to beam scintillations, wanderings as a result of random fluctuations of atmospheric temperature and refractive index changes, humidity, pressure and their interactions. A scintillometer measures the normalized variance of radiation intensity.

The scintillometer set up consists of an emitter and a receiver, placed in front of each other at the distance L, where the measurement is made.

The scintillometer consists in an emitter and a receiver. The emitter includes a continuous laser and a pair of lenses to collimate the beam over the optical path. The laser wavelength used is 532 nm with an average power of 70 mW. The output beam of the laser beam is expanding with a Galilean telescope with chosen lenses. The beam is then collimated over a long distance typically for distances less than 200 m. The optical system has a plano-convex lens with a focal length of 15 mm after the laser output. In order to have a magnification of 5 a second plano-convex lens with a 300 mm, focal length is placed 30 cm from the first one. Finally, at the output of the emitter, the diameter of the beam is then 2 cm. A Rayleigh range of around 142 m for the output collimated beam is obtained with the pair of lenses. The receiver uses a Galilean telescope to recollect the light. A position sensing detector sensor with lateral effect (Duo lateral PSD) defined by the size of the light spot is used as the detector and provide position informations only up to the point where the edge of the spot reaches the gap. The lateral effect position sensing detector is 100 mm^2 (10 mm x 10 mm). Additionally, it is important to enlarge the beam with a combination of two lenses enabling a magnification of 5 in order to have a sufficiently large measurement field. The first lens is plano-convex with a focal length of 60 mm, and the second is plano-concave with a focal length of −24 mm. The two lenses are separated with a distance of 86 mm. For stability during the measurements, all the optical components are mounted on a metal board. The electronic system includes an electronic system for data acquisition and a remote interface with the operators. The scintillometer is a stand-alone system with batteries, solar cells and a communicating system to send data. That device has been built for simultaneously recording both random intensity fluctuations and displacement of the beam centroid (wandering).

The scintillometer provides a measurement of the structure parameter for the refractive index (C_n^2) along the optical path. The structure parameter of the refractive index of air C_n^2 was calculated from the natural logarithm of the intensity of light (I)

$$\sigma_{lnA}^2 = 0.031k^{7/6}L^{11/6}C_n^2 \quad (1)$$

where L is the beam path length (m), k is the optical wave number (m^{-1}) defined for the wavelength λ as: $k = 2\pi/\lambda$ and $\sigma_{lnA}^2 = \frac{1}{4}ln(1+\sigma_I^2)$ with σ_I^2 defined as: $\sigma_I^2 = \langle I^2\rangle - \langle I\rangle^2/\langle I\rangle^2$. So the representative value of C_n^2 is 10^{-15} to $10^{-18}m^{-2/3}$.

The parameters such as temperature (T), humidity (q) and pressure (P) generate fluctuations in the refractive index of air C_n^2; however, as the proportion of pressure is very small, this value is always neglected. In the range of visible and near-infrared region, the temperature is the main parameter, assuming that temperature and humidity are perfectly correlated, the structure parameter C_n^2 can be related to the structure parameter C_T^2 for temperatures by [12]

$$C_T^2 = C_n^2\left(\frac{0.78\times10^{-6}P}{T^2}\right)^{-2}\left(1+\frac{0.03}{\beta}\right)^{-2} \quad (2)$$

where β is the Bowen ratio, which connect temperature and humidity by the ratio of sensible flux and latent heat flux ($\beta = H/\lambda E$). The second term in brackets is a correction for the effects of humidity. C_T^2 is given in ($K^2.m^{-2/3}$).

The sensible heat flux can be derived from the Monin-Obukhov Similarity Theory (MOST) [13] when C_T^2 is known. This value depends on the stability parameter $\zeta = (z_{scin} - d)/L$, where z_{scin} and d are the effective height of the scintillometer above the surface and the displacement height, respectively. L_O is the Monin-Obukhov length (m) given by

$$L_O = \frac{u^2 T}{kgT^*} \tag{3}$$

with $k = 0.41$ is the von Karman constant, $g = 9.81\ ms^{-2}$ the gravity and $u^* \left[m.s^{-1}\right]$ the friction velocity, given by

$$u^* = \frac{ku}{ln\left(\frac{z-d}{z_0}\right) - \psi} \tag{4}$$

where u is the wind speed, $z_0 = 0.1h_{veg}$ is the surface roughness length and ψ is the stability correction function depending on z/L_O. The universal function ψ is only related to the atmosphere stability and has different expression under stable and unstable conditions [14]. The sensible heat flux H_{scin} can be then computed iteratively as follows

$$H_{scin} = -\rho c_p u^* T^* \tag{5}$$

where ρ and c_p are the density and specific heat capacity of the air, respectively. During the iteration, β is calculated using H_{scin}, net radiation (R_n) and soil heat flux (G)

$$\beta = \frac{H_{scin}}{R_n - G - H_{scin}} \tag{6}$$

The value of the latent heat flux (evapotranspiration) can then be calculated as the residual of the energy balance

$$\lambda_v E_{scin} = R_n - G - H_{scin} \tag{7}$$

where R_n $\left[W.m^{-2}\right]$ is the net radiation and G $\left[W.m^{-2}\right]$ is the soil heat flux. Additional data of temperature, pressure and humidity are necessary to compute the characteristic parameters of the latent heat flux. More specific information on the described approach is found in Ref. [15].

Simultaneously, the air and ground temperature were measured with a thermistor device and linked to the transmission system. The humidity was measured with a capacitive sensor and the wind speed with an anemometer. All the data were collected by a data acquisition electronic system, based on an Arduino board [16], and sent as text files by a modem through the GPRS network and a SIM card.

2.3. Estimation of evapotranspiration based on micro-meteorological datasets

In addition to the data provided by the scintillometer, estimation of the reference crop evapotranspiration ET_0 can be based on energy balance schemes and the Penman-Monteith

(FAO-PM56) method [17, 18]; they are used to assess ET_0 from meteorological variables. The reference crop evapotranspiration or reference evapotranspiration, denoted as ET_0 or ET_{ref}, is the estimation of the evapotranspiration from a "reference surface." The reference surface is a hypothetical grass reference crop with an assumed crop height of 0.12 m, a fixed surface resistance of 70 S/m and an albedo of 0.23. In the reference evapotranspiration definition, the reference surface is specifically defined as the reference crop, and this crop is assumed to be free of water stress and diseases with a fixed surface resistance of 70 S/m. This reference surface has grass with a uniform height, normally growing and totally covering the ground. The soil surface is moderately dry resulting from about a weekly irrigation frequency.

The FAO Penman-Monteith method has been reported as providing consistent ET_0 values in many regions and climates [19]; it has long been accepted worldwide as an excellent ET_0 estimator when it is compared with other methods. The application of ET_0 models with fewer meteorological variable requirements is recommended under situations where weather data sets are incomplete. Those values are multiplied by an empirical crop coefficient to obtain the ET from the crop (ET_c). The crop coefficient accounts for the difference between the standard surface and the crop. Reference ET is expressed in units of depth time^{-1}, for example, mm day^{-1}. It is a climatic parameter expressing the evaporative power of the atmosphere at the given space and time coordinates [20].

Empirical formulas have been developed to estimate solar radiation using some normal observations from meteorological stations, such as maximum and minimum temperatures, sunshine hours, cloud, precipitation, latitude and elevation. FAO Penman-Monteith procedures allow applying the method when only air temperature and wind speed are available. In these methods, saturation vapor pressure and actual vapor pressure were estimated from T_{max} and T_{min}, as recommended by Allen et al. [22] for situations where air humidity data are lacking or are of questionable quality.

The standard FAO Penman-Monteith method is based on the following equation [21]

$$ET_0 = \frac{0.408\Delta(R_n - G) + \gamma\left(\frac{0.37}{(T_a + 273)}\right)u_2(e_s - e_a)}{\Delta + \gamma(1 + 0.34u_2)} \quad (8)$$

where ET_0 represents the hourly reference evapotranspiration (mm h^{-1}), Δ represents the slope vapor pressure curve (kPa.°C^{-1}), R_n is the net radiation at the crop surface (MJ m^{-2} h^{-1}), G indicates the soil heat flux density (MJ.m^{-2}.h^{-1}), γ is the psychrometric constant (kPa.°C^{-1}), T_a is the mean air temperature at 2 m height (°C), u_2 is the mean hourly wind speed at 2 m height (m s^{-1}), e_s defines the saturation vapor pressure at air temperature T_a (kPa) and e_a indicates the actual vapor pressure (kPa). All meteorological data required by the equation were collected by the aid of a weather station that was placed on the reference crop.

The Penman-Monteith equation requires the following parameters:

a. Minimum and maximum daily temperature.

b. *Relative humidity.* Depending on the availability of data, different equations are used. The data requirements are the following: (i) minimum and maximum daily relative humidity, (ii) maximum daily relative humidity or (iii) average daily relative humidity. In case where no humidity data are available, an estimation is required considering that the dew point temperatures are the same as the daily minimum temperatures.

c. *Solar radiation.* Different equations are used to consider the solar radiation. In order to calculate the solar radiation, if we do not access to the solar radiation directly we need (i) hours of sunshine per day or (ii) cloudiness fraction or the Hargreaves formula, based on minimum and maximum daily temperature and an adjustment coefficient (Krs) to estimate the solar radiation.

d. Wind speed. (An adjustment can be made if the wind speed measurement height has been previously measured).

e. Latitude and altitude of the climate measurement station.

f. The meteorological data measured and used in this study are mean daily air temperature (T_{mean}); maximum and minimum air temperature (T_{max} and T_{min}); mean daily relative humidity (RH); mean daily wind speed (u) and daily net radiation (R_n).

3. Results and discussion

3.1. Comparison between micrometeorological measurements and scintillometry

The comparison period for the ET measured by scintillometry and micro-meteorology measurements covered 2 months between July and August 2015, which included a range of climatic conditions to test the performance of the proposed scintillometry calculation method. The average air temperature was 23°C and ranged between a minimum of 9°C and a maximum of 32°C. The average ground temperature was 22°C and ranged between a minimum of 16°C and a maximum of 25°C. The C_n^2 measurements were made on a continuous time scale at 60 min intervals and averaged every 1 min and were synchronized with a weather station. R_n and G are calculated based on other meteorological data. Sensible heat fluxes with 60 min intervals were calculated using a step by step methodology provided by the method previously developed. The 60 min ET rates were computed as a residual from the energy balance equation, with calculated R_n and G fluxes and using an estimated H. **Figures 2** and **3** show the data acquired by the micro-meteorological station. **Figure 2** displays air and ground temperature measurements and, besides, the wind speed.

Figure 3 presents the measurements of humidity and pressure. All the data were acquired, at the same time, and stored by the electronic system. The measurements are detailed on a period of 7 days for simplicity of analysis. The air temperature was measured at a height of 0.3 m above the crop canopy and shows large variations between the maxima and minima. We calculate a ΔT around 15°C. The ground temperature was measured beneath the canopy foliage

http://dx.doi.org/10.5772/intechopen.68538

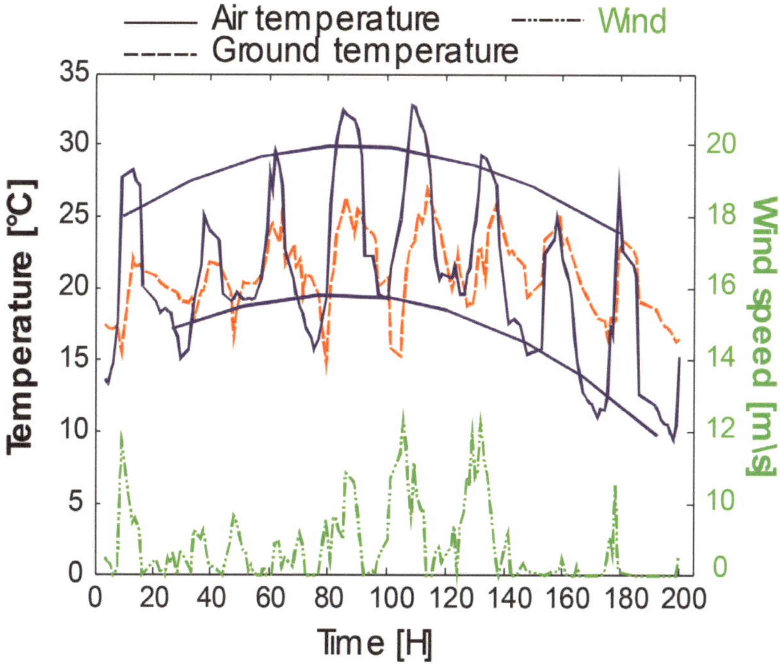

Figure 2. Measurements of air, ground temperature and wind during considered period. The lines give the average minimal and maximal air temperature.

with a sensor at 0.05 m underground. The ground temperature data show small amplitudes of variation with an average ΔT between the maximum and minimum of temperature around 5°C. The comparison between the two measured temperatures at two different heights shows a constant average value for the ground temperature and a limited variation of the average temperature with a maximum value of 25°C. It is interesting to note that the maximum of average air temperature value corresponds to the days where the wind is the maximum. The ground temperature is not sensible to the wind and account for the temperature changes with the canopy as shelter. The maximum in the average value for the air temperature corresponds also to the maximum of humidity. However, the atmospheric pressure is not related to the air temperature during the presented period of analysis. Based on the acquired micro-meteorological parameters and atmospheric turbulence measurements, evapotranspiration was calculated.

Figure 4 presents the comparison between the evapotranspiration data acquired with the scintillometer and the FAO-56 method during the same period of atmospheric parameter measurements. Differences are observed, in particular, during the night where the temperature falls. Those differences can be attributed to the few numbers of parameters used in the calculation with the FAO-56 method. The discrepancies between the two measurements of ET correspond to the maximum of humidity and minimum of wind. The variation between the two measurements can be checked in **Figure 5**, where the ETP computed using the FAO method regarding the ETP calculated with scintillometer data are plotted. A coefficient of

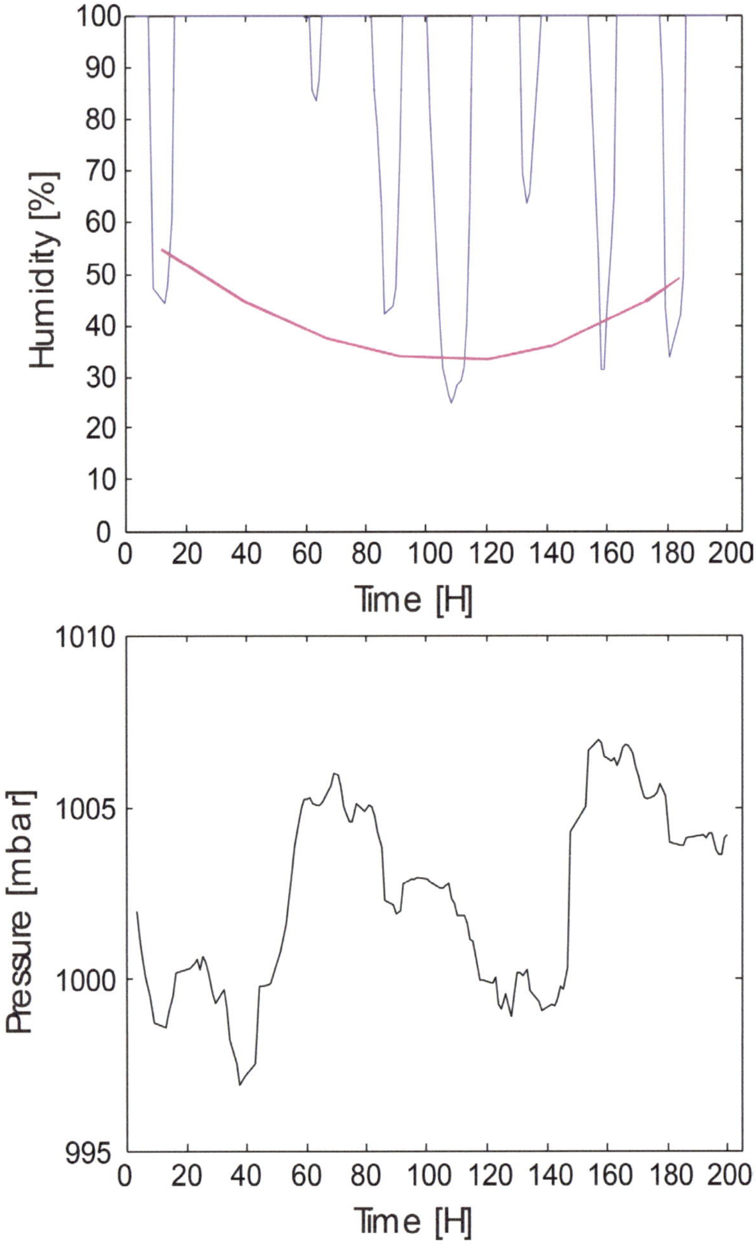

Figure 3. Measurement of the atmospheric humidity and pressure.

0.78 is found between the two methods meaning that $ETP_{FAO} = 0.78\ ETP_{scin}$. Therefore, the values are lower with FAO than with scintillometry and, consequently, evapotranspiration is underestimated. Note that the evapotranspiration measurement with scintillometry has proven being close to the data measured with Eddy-covariance [22, 23].

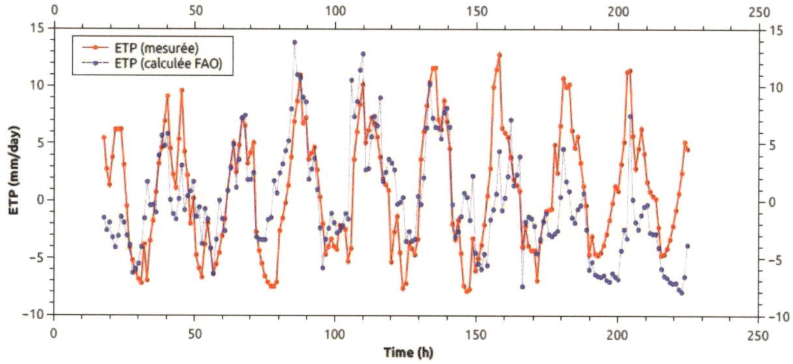

Figure 4. Comparison of the ET measured with the FAO-PM56 model based on the measurements of a limited number of parameters (T_{min}, T_{max}, T_{avg}, R_s) and scintillometric data.

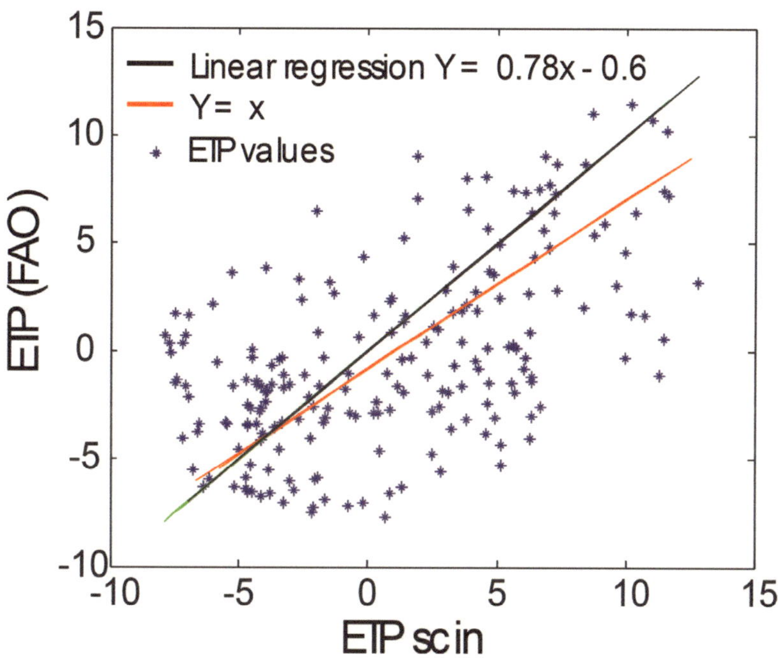

Figure 5. Comparison of the two measured ET data. The curves show the fitting points calculated with linear regression and the ideal curve representing a perfect similar relation of the two ET.

3.2. Influence of environmental parameters

In order to account the influence of the main meteorological parameters on the measurement sensibility of ETP, the effect of temperature variations is observed as the main influence on the measurements obtained by the FAO-56 and the scintillometer. Thus, the calculation is limited to temperature differences since scintillometer is more sensitive to small fluctuations of the refractive index of the air caused by those variations of temperatures.

Firstly, considering the ETP solution from a set of variable values (Eq. (8)) given by the FAO-56 where the dominant variable is the air temperature (T_A). Indeed, each term in Eq. (8) can be rewritten as a function of the temperature, thus

$$ET_0 = \frac{[0.408.\Delta R_n + \gamma.\frac{900}{(T_A+273)} .u_2.(e_s - e_s.H_D)]}{(\Delta + \gamma.\ (1 + 0.34.u_2))} \tag{9}$$

where u_2 is the wind speed and H_D is the relative humidity. R_n, e_s, γ and Δ are depending on T_A (see Appendix A).

The dominant variable value is fluctuated by a small amount while keeping all other values constant, and we note the change of the solution. The goal is to determine how sensitive the output calculation of evapotranspiration could be with respect to the calculation elements which are subject to uncertainty of variability.

Figure 6 shows that a variation of ±0.5°C of T_A leads to a small variation of ET_0 (±1.3 mm/day) with the FAO method. However, a variation of ±0.5°C leads to a variation of ±2.2 mm/day with the scintillometric method. Therefore, the scintillometer method is more sensitive to fluctuations of air temperature. This is not a novelty because small fluctuations of air temperature induce random variations in the refractive index of the turbulent atmosphere by changing the intensity of turbulence C_n^2. The propagation distance enhances any fluctuations of the laser beam. Moreover, it was shown that the C_n^2 can be measured with a good precision [24]. According to this result, a sensitivity analysis based on the analytical calculations was conducted.

The approach consists in mathematically differentiate the equation under study to derive equations for the change rate of the independent variable with respect to each dependent variable. The sensitivity for the air temperature is calculated applying the PM FAO-56 method. The ET_0 is considered as a function with multi-variables v_1, v_2, v_3, …, as $ET_0 = f(v_1, v_2, v_3, \ldots)$ with $v_1 = T_a$, $v_2 = R_n$, $v_3 = u_2$ and so on. The sensitivity equation can be developed by

$$ET_0 + \Delta ET_0 = f(v_1 + \Delta v_1, v_2 + \Delta v_2,\ v_3 + \Delta v_3,\ \ldots)\ . \tag{10}$$

After applying the Taylor's theorem and considering only the first order, the expression yields

$$\Delta ET_0 = \frac{\partial ET_0}{\partial v_1}\Delta v_1 + \frac{\partial ET_0}{\partial v_2}\Delta v_2 + \frac{\partial ET_0}{\partial v_3}\Delta v_3 + \ldots \tag{11}$$

The P–M method is a multi-variable model. Several variables have different dimensions and various ranges of values, which makes it difficult to compare the sensitivity by partial derivatives. The partial derivative is transformed into a non-dimensional form.

The substitution of the relative forms, $(ET_0)_{rel} = (\Delta ET_0/ET_0)$ and $v_{rel} = \Delta v/v$ for each variable, yields

$$(ET_0)_{rel} = (\Delta ET_0/ET_0) = \left(\frac{\partial ET_0}{\partial v_1}\frac{v_1}{ET_0}\right)\frac{\Delta v_1}{v_1} + \left(\frac{\partial ET_0}{\partial v_2}\frac{v_2}{ET_0}\right)\frac{\Delta v_2}{v_2} + \left(\frac{\partial ET_0}{\partial v_3}\frac{v_3}{ET_0}\right)\frac{\Delta v_3}{v_3} + \ldots \tag{12}$$

http://dx.doi.org/10.5772/intechopen.68538

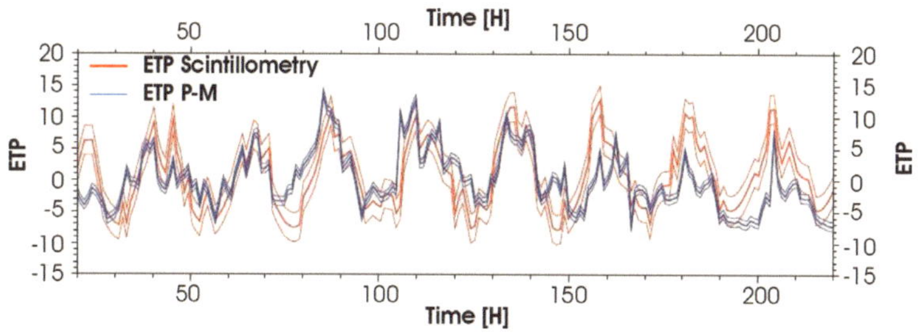

Figure 6. Comparison of the influence of a small variation of temperature (± 0.5°C) on the two methods of measurements of ET.

where Sv_i is called the sensitivity coefficient and it is equal to $\frac{\partial ET_0}{\partial v_i}\frac{v_i}{ET_0}$ for the variable v_i. This term becomes a dimensionless coefficient which expresses the percentage of the relative variable change transmitted to the relative dependent variable. Basically, a positive/negative sensitivity coefficient of a variable indicates that ET_0 will increase/decrease as the variable increases. The bigger the sensitivity coefficient, the larger the effect a given variable has on ET_0. A sensitivity coefficient Sv_i of 0.2 would show that a 10% change in v1 ($\Delta v_1/v_1 = 0.10$) would cause a 2% change in ET_0 ($\Delta ET_0/ET_0 = 0.02$) if $(ET_0)_{rel}$ is dependent of the relative change $\Delta V_1/V_1$ in Eq. (12) [25].

The partial derivatives needed for the determination of the sensitivity coefficient S_{TA} corresponding to the influence of air temperature:

$$S_{TA} = \lim_{\Delta T_A \to 0}\left(\frac{\frac{\Delta ET_0}{ET_0}}{\frac{\Delta T_A}{T_A}}\right) = \frac{\partial ET_0}{\partial T_A}\cdot\frac{T_A}{ET_0} \tag{13}$$

The calculation is done analytically by means of symbolic calculation of Mathematica (Wolfram) in Appendix A. It found a sensitivity coefficient value of 0.12 for the maximum measured temperature of 32.8°C, a humidity of 24.7%, a wind measurement of 3.9 m/s, an evapotranspiration of 10.4 mm/day and an atmospheric pressure of 1002 mbar.

This sensitivity obtained for the FAO-56 measurement and the one given by the scintillometer is compared. For the latter instrument, based on optical metrology, the intensity fluctuations of visible beams are more sensitive to temperature fluctuations than humidity fluctuations.

In the displaced-beam scintillometer measurements, the path-averaged measurements of C_n^2 are obtained. Additional measurements have separately carried out including temporally averaged pressure, air temperature, humidity, as well as the height of the beam above the field and the Bowen ratio. All those sources of measurements contain uncertainties. Uncertainty is propagated from the measured parameters to the derived variables through the set of equations employed and written previously (Eq. (1)–(7)). Different scintillometer sensitivity studies have been done. One of them uses the Monte-Carlo error analysis [26] and shows that the experimental coupling of inertial-dissipation methods is promising, since the propagation of statistical errors in the acquired parameters to the final value is limited. Those methods are based on measurements of the structure parameters of momentum, temperature and humidity with optical methods leading to the calculation of the momentum flux and the heat flux. Another study [27] calculates the relative uncertainty of the friction velocity u* or turbulence velocity scale. The goal was to know how precisely the measurement of the friction velocity u* can be done by using a path-averaged optical propagation. The conclusion is that the measurement of the inner scale of turbulence (λ_o) contributes to the largest uncertainty and must be done precisely. A recent study analyses the impact of the Bowen ratio on the flux value and uncertainty [28]. It is shown that the Bowen ratio has a large impact on the accuracy of C_T^2 and on the sensible heat flux estimation in the case of strong humidity conditions ($\beta < 1$). A $\beta > 1$ was registered during the summer experiment. A relative uncertainty is estimated on the measurements, considering only the temperature with the scintillometer and following the procedure described for remote sensing [29]

$$\frac{\Delta ET_0}{ET_0} = \left(\frac{\Delta T_A}{T_A}\right) S_{T_A}. \tag{14}$$

The following tolerances used to estimate uncertainties have been taken:

Quantities	Unit	Assumed standard deviation
Temperature	°C	±1°C (−20–0°C)
		± 0.5°C (0–40°C)
		± 1°C (40–60°C)
Wind speed	$m.s^{-1}$	± 0.5 m/s (0–20 m/s)
		± 3% (20–60 m/s)
Humidity	%	± 2% (0–90%)
		± 3% (90–99%)
		Temperature dependence
		± 0.05% RH°C^{-1}
Atmosphere pressure	Pa	± 50 Pa (300–1100 hPa)
Path length	m	± 3
Path height	m	0.2
C_n^2	$K^2.m^{-2/3}$	± 0.5%

A value of 0.05 is found for the scintillometer sensitivity corresponding to sensitivity 2.4 lower than the FAO-56 sensitivity.

The main factor of uncertainty comes from the measurement of C_n^2. The scintillometer uncertainty is lower than the FAO-56 uncertainty.

3.3. Neural network for estimating evapotranspiration

Several researchers have used artificial neural network (ANN) models to estimate or forecast evapotranspiration as a function of micro-meteorological data [30, 31]. Neural networks are an information processing technique based on a biologically-inspired programming paradigm, such as the brain enabling computer softwares to learn from observational data. The similarity with the human brain consists in the following characteristics: (i) a neural network acquires knowledge or informations through a process of learning; (ii) a neural network's knowledge is stored within inter-neuron connection strengths known as synaptic weights. In a feed-forward ANN model, a neuron performs two functions; it sums the weighted inputs from several connections and then applies a nonlinear function to the sum. The resulting value is propagated through outgoing connections to other neurons. The neurons or interconnection points are arranged in layers. The input layer receives data as inputs from real data acquisition; succeeding layers receive weighted outputs from the preceding layer as inputs, and the last layer gives the final results. ANNs are trained using a training algorithm and a training data set to adjust the connection weights, which result in an ANN model that can generate the most similar output vector to the target vector [32]. It is important to note that in most of the papers, different ANN models are considered including the generalized feedforward (GFF), linear regression (LR), multi-layer perceptron (MLP) and probabilistic neural network (PNN) [33].

The most common neural network model is the multi-layer perceptron (MLP). The MLP and many other neural networks learn using an algorithm called backpropagation artificial neural networks (BPANN). With backpropagation, the input data are repeatedly presented to the neural network. As soon as the calculation is done, the obtained result at the final layer of the neural network is compared to the desired output, and an error is calculated. This error is then fed back (backpropagated) to the neural network input layer and used to adjust the synaptic weights such that the error decreases with each iteration. Finally, the neural model of synaptic connections converges closer and closer to the desired output. These calculations with feedbacks are known as "training."

The objective of this study was to test BPANN models for forecasting daily ET_O with input data based on minimum meteorological data, considering the mean for maximum and minimum air temperatures and extraterrestrial radiation for the FAO-PM56 and scintillometry model for finally to compare them. The comparisons were based on statistical differences during a period of measurements with forecasted data by a set of measured data as input parameters and compared to data measured using FAO-PM56 or scintillometry daily ET_O values as references.

Firstly, the FAO-PM56 method was used to calculate daily ET values from climatic data. The ET data were then used to train and test the ANN model. 120 h are considered corresponding to 5 days as training data. Secondly, evapotranspiration data based on the scintillometer data were used to train and test the ANNs model. The same number of days is kept as input data.

Figure 7(a) shows the plot of the predicted data as a function of the measured data for scintillometry. A shift in the plot of the predicted values for the input data acquired with the scintillometry method is seen. However, the predicted data follow the measured data variations. The slope of the fitting curve is 0.76 close to the perfect value of 1 in the case of similar values between the predicted and measured data (**Figure 7c**). With FAO-PM-56

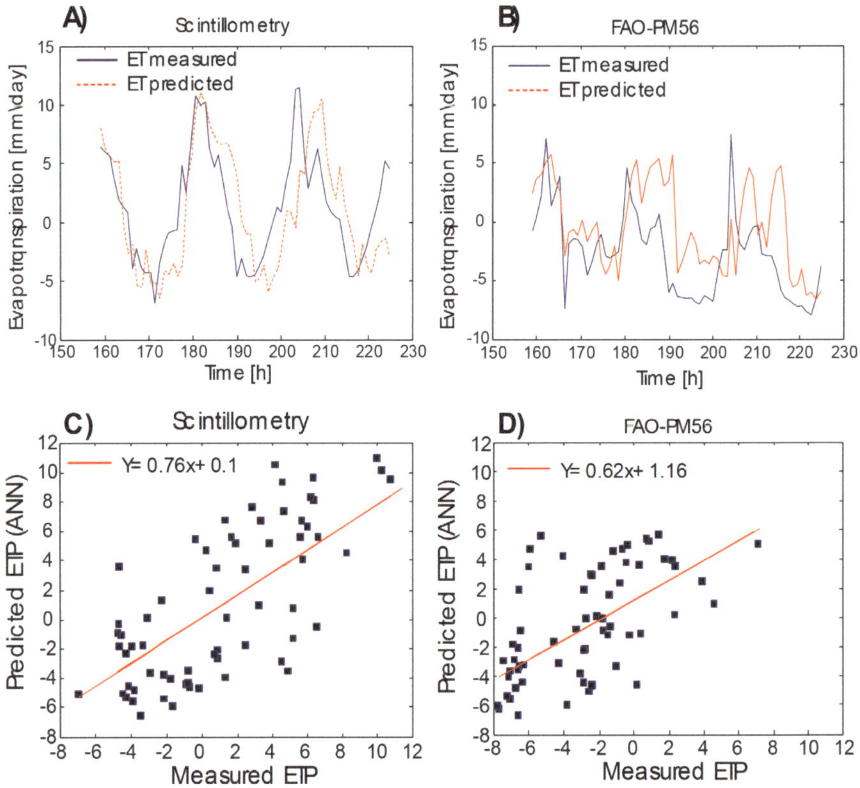

Figure 7. Predicted values of ET based on two different input training data of ET one with FAO-PM56 and the other one with scintillometric data. The calculation is done with an artificial neural network (ANN).

data, the BPANN calculations show that the predicted values have no time shift but large differences in intensity (**Figure 7b**). (**Figure 7d**) presents the corresponding plot of the predicted values as a function of the measured values. The slope is 0.62 meaning that the predicted values are almost close to twice the measured values. The comparison between the two predicted values is presented in **Figure 8** where the plot of the two calculations with the ANN model shows a slope of 1.5 demonstrating the big difference in the obtained values of ET. As a conclusion, the scintillometry data are better input values for the forecasting of ET.

In order to know if the difference in the final results of the predicted values done with the ANN model comes from the capacity to sufficiently have training data, the difference between the predicted and measured values as a function of the number of days of input data is calculated. **Figure 9** displays the optimum values. It is observed that up to 2.5 days of forecasting, there is no difference between the two methods of evapotranspiration calculations. Nevertheless, after 2.5 days, the evapotranspiration calculation with the scintillometer shows constant values in the difference between measured and predicted scintillometric input data. This is not the case with FAO-PM56 where the difference between predicted and measured data increases. The scintillometry is more able to predict for more days than the FAO-PM56

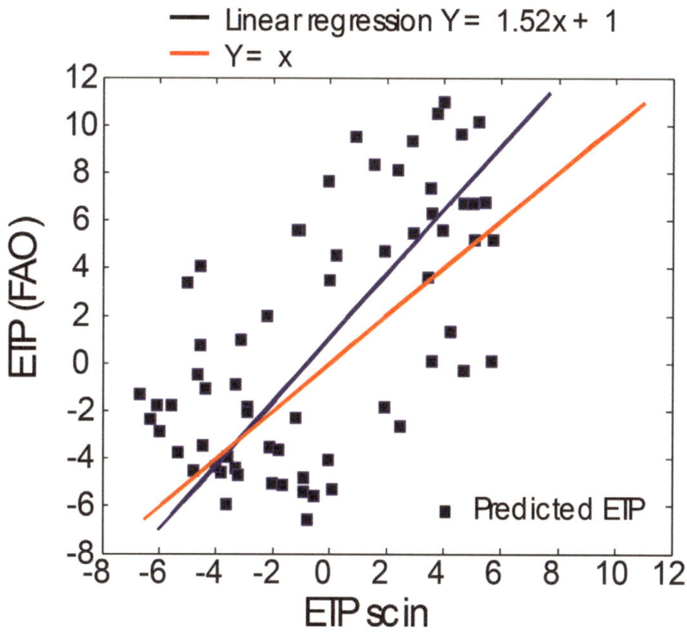

Figure 8. Comparison of the two sets of forecasted values with an artificial neural network. The curves show the fitting points and the ideal case of similar values.

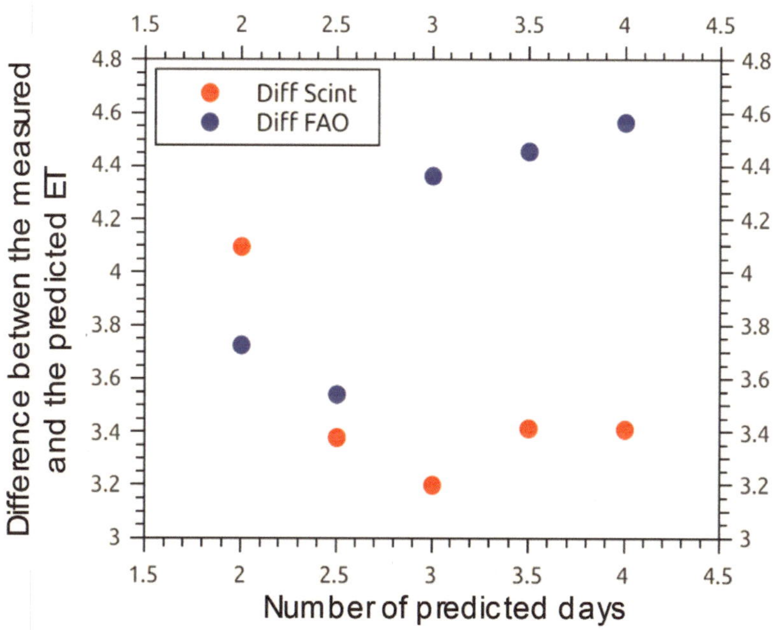

Figure 9. Calculation of optimal forecasted days as a function of the number of input training days in the ANN calculation for two different methods of ET measurements.

method. This result can be explained by the low propagation of small variations or errors in the calculation of ET using the scintillometer method. However, in the FAO-PM56 method with few input variables, small variations in the parameters for the calculation of ET lead to larger uncertainties.

4. Conclusion

Two methods to acquire ET from a crop field (Maize) are presented. One of them is based on atmospheric turbulence data acquired with a laser beam scintillometric method and another one is based on the calculation with limited meteorological data using the FAO-PM56 method. The comparison between the two ET measurements shows a difference in the final result. The measurements with micro-meteorological parameters are lower than with scintillometric parameters leading to an underestimation of the real ET. This is an important result because farmers must accommodate in advance their crop water demand to irrigation requirements. An overestimation of ET can lead to a deficit of irrigation water, and on the contrary, a low estimation of ET can lead to water waste. The measurements have shown a lower measured value of ET with the FAO-PM56 method. Moreover, the FAO-PM56 method for obtaining ET is more sensible to a small error in the acquisition of the temperature. Then, while the scintillometer measurements are representative of the turbulent fluctuations along the whole beam path, the FAO-PM56 measurements are typically representative measurements of localized areas near the respective different meteorological acquisition instruments.

Finally, using artificial neural network, evapotranspiration forecast for short-term near-future is obtained. In addition, it is presented results for two different input training data, and it is showed that evapotranspiration data based on scintillometric data acquisition are more reliable for forecasting. An optimum value was found in the number of days of training data to obtain the best forecast. In this case also, evapotranspiration with scintillometric data increases the number of predicting reliable days.

Appendix A. Supplementary data

ETP solution from a set of variable values with a Mathematica script:

```
et[TA_, WD_, HD_, RS_, H_, DAY_, Y_, LAT_] :
= (0.408*(4098*( 0.611*Exp[17.27*TA/(TA + 237.3)])/(TA
  + 237.3)^2)*((0.77*RS) − ((4.903*10^(−9)*(TA + 273)^4
  *(0.34 − 0.14*Sqrt[( 0.611*Exp[17.27*TA/(TA + 237.3)])*HD]))
  *(1.35*(Min[RS/((0.75 + 0.00002*H)
  *(24*60/Pi*(0.0820)*(1 + 0.033*Cos[2*Pi * DAY/Y])
  *((\[Pi]/2 − ArcTan[(−Tan[(3.14/180*LAT)]*Tan[(0.409*Sin[2*\[Pi]
  *DAY/Y − 1.39])])/(Max[1 − Tan[(3.14/180*LAT)]^2*
```

Tan[(0.409*Sin[2*Pi * DAY/Y − 1.39])]^2 , 0.00001])^0.5])
*Sin[(3.14/180*LAT)]*Sin[(0.409*Sin[2*Pi * DAY/Y − 1.39])]
+Cos[(3.14/180*LAT)]*Cos[(0.409*Sin[2*Pi * DAY/Y − 1.39])]
Sin[(Pi/2 − ArcTan[(−Tan[(3.14/180*LAT)]*Tan[(0.409
*Sin[2*Pi * DAY/Y − 1.39])])/(Max[1 − Tan[(3.14/180*LAT)]^2
*Tan[(0.409*Sin[2*Pi * DAY/Y − 1.39])]^2 , 0.00001])^0.5])]))), 1]) − 0.35)))
+(0.00163*(101.3*((TA + 273 − 0.0065*H)/(TA + 273))^5.26)/
(2.501 − (0.002361)*TA))*900/(TA + 273)*WD*((0.611
*Exp[17.27*TA/(TA + 237.3)]) − (0.611*Exp[17.27
*TA/(TA + 237.3)])*HD))/((4098*(0.611*Exp[17.27*TA/(TA
+237.3)])/(TA + 237.3)^2) + (0.00163*(101.3*((TA + 273 − 0.0065*H)/
(TA + 273))^5.26)/(2.501 − (0.002361)*TA))
*(1 + 0.34*WD));

Author details

Antonin Poisson[1], Angel Fernandez[1,3], Dario G. Gomez[2], Régis Barillé[1]* and Benoit Chorro[4]

*Address all correspondence to: regis.barille@univ-angers.fr

1 Moltech-Anjou, University of Angers, Angers, France

2 Instituto de Física, Pontificia Universidad Católica de Valparaíso, Valparaíso, Chile

3 Departamento de Física, Universidad Técnica Federico Santa María, Valparaíso, Chile

4 Océalia, CIVRAY, France

References

[1] Intergovernmental Panel on Climate Change. Synthesis Report. 2014. https://www.ipcc.ch/report/ar5/

[2] Dasa B, Singha A, Pandaa S N, Yasuda H. Optimal land and water resources allocation policies for sustainable irrigated agriculture. Land Use Policy. 2015;**42**:527–537

[3] Abtew W, Melesse A. Evaporation and Evapotranspiration. Berlin, Heidelberg: Springer-Verlag; 2013

[4] Béziata P, Rivallanda V, Tallec T, Jarosza N, Bouleta G, Gentine P, Ceschia E. Evaluation of a simple approach for crop evapotranspiration partitioning and analysis of the water budget distribution for several crop species. Agricultural and Forest Meteorology. 2013;**177**:46–56

[5] Pereira L, Allen R G, Smith M, Raes D. Crop evapotranspiration estimation with FAO56: Past and future. Agricultural Water Management. 2013;**147**:4–20

[6] Verstraeten W W, Veroustraete F, Feyen J. Assessment of evapotranspiration and soil moisture content across different scales of observations. Sensors. 2008;**8**:70–117

[7] Beyrich F, Bange J, Hartogensis O K, Raasch S, Braam M, van Dinther D, Gräf D, van Kesteren B, van den Kroonenberg A C, Maronga B, Martin S, Moene AF. Towards a Validation of Scintillometer Measurements: The LITFASS-2009 Experiment. Boundary Layer and Mesoscale Meteorology. 2012;**144**:83–112

[8] Meireles MRG, Almeida PEM, Simões MG. A Comprehensive Review for Industrial Applicability of Artificial Neural Networks, IEEE Trans. Indust. Elec. 2003;**50**(3): 585–601

[9] Gardner MW, Drorling SR. Artificial neural networks (the multilayer perceptron)—A review of applications in the atmospheric sciences. Atmospheric Environment. 1998;**32** (14/15):2627–2636

[10] Kleissl J, Gomez J, Hong SH, Hendrickx JMH, Rahn T, Defoor WL. Large Aperture Scintillometer Intercomparison Study. Boundary Layer and Mesoscale Meteorology. 2008;**128**(1):133–150

[11] Chehbouni A. Estimation of heat and momentum fluxes over complex terrain using a large aperture scintillometer. Agricultural and Forest Meteorology. 2000;**105**:215–226

[12] Hill RJ. Algorithms for obtaining atmospheric surface-layer fluxes from scintillation measurements. Journal of Atmospheric and Oceanic Technology. 1997;**14**:456-467

[13] Cammalleri C, Agnese C, Ciraolo G, Minacapilli M, Provenzano G, Rallo G. Actual evapotranspiration assessment by means of a coupled energy/hydrologic balance model: Validation over an olive grove by means of scintillometry and measurements of soil water contents. Journal of Hydrology. 2010;**392**:70–82

[14] Zhang X, Jia X, Yang J, Hua L. Evaluation of MOST functions and roughness length parameterization on sensible heat flux measured by large aperture scintillometer over a corn field. Agriculture and Forest Meteorology. 2010;150:1182–1191

[15] Odhiambo GO, Savage MJ. Surface layer scintillometry for estimating the sensible heat flux component of the surface energy balance. South African Journal of Science. 2009;**105**:208–216

[16] de Bruin HAR. Analytic solutions of the equations governing the temperature fluctuation method. Boundary Layer and Mesoscale Meteorology. 1994;**68**(4):427–432

[17] Subedi A, Chávez J L, Crop Evapotranspiration (ET) Estimation Models: A Review and Discussion of the Applicability and Limitations of ET Methods, Journal of Agricultural Science. 2015;**7**(6): 50–68.

[18] Novák V. Methods of Evapotranspiration Estimation', Evapotranspiration in the Soil-Plant-Atmosphere System, Progress in Soil Science. 2012; editor Springer Netherlands, 165–215

[19] Rivasa R, Caselles V: A simplified equation to estimate spatial reference evaporation from remote sensing-based surface temperature and local meteorological data. Remote Sensing of Environment. 2004; **93**:68–76

[20] Droogers P, Allen R G: Estimating reference evapotranspiration under inaccurate data conditions, Irrigation and Drainage Systems. 2002; **16**:33–45

[21] McMahon T A, Peel M C, Lowe L, Srikanthan R, McVicar T R: Estimating actual, potential, reference crop and pan evaporation using standard meteorological data: A pragmatic synthesis, Hydrology and Earth System Sciences Journal. 2013; **17**:1331–1363

[22] Allen R G, Pereira L S, Raes D, Smith M: Crop evapotranspiration: guide lines for computing crop water requirements, FAO irrigation and Drainage, Paper No. 56. 1998; FAO: Rome, Italy

[23] Hoedjes J C B, Chehbouni A, Ezzahar J, Escadafal R, De Bruin H A R: Comparison of Large Aperture Scintillometer and Eddy Covariance Measurements: Can Thermal Infrared Data be used to capture footprint-induced differences, Jopurnal of Hydrometeor. 2007; **8**:144–158

[24] Wang T, Ochs G R, and Clifford S F: A saturation-resistant optical scintillometer to measure C_n^2, JOSA. 1978; **68**:334–338

[25] Liu SM, Xu ZW, Zhu Z L, Jia Z Z, Zhu M J, Measurements of evapotranspiration from eddy-covariance systems and large aperture scintillometers in the Hai River Basin, China, Journal of Hydrology. 2013; **487**:24–38

[26] Saxton K E: Sensitivity analyses of the combination evapotranspiration equation, Agricultural Meteorology. 1975; **15**:343–353

[27] Moroni C, Navarra A, Guzzi R: Estimation of the turbulent fluxes in the surface layer using the inertial dissipative method: A monte-carlo error analysis, Boundary Layer and Mesoscale Meteorology. 1990; 50:339–354

[28] Andreas E L: Uncertainty in a path-averaged measurement of the Friction Velocity u*, Journal of Applied Meteorology and Climatology. 1992; **31**:1312–1321

[29] Ustinov E A: Sensitivity Analysis in Remote Sensing, Springer Briefs in Earth Sciences, Springer. 2015

[30] Govindaraju R S, Rao A R: Artificial Neural Networks in Hydrology, Kluwer Academic Publishers, Amsterdam. 2000

[31] Traore S, Luo Y, Fipps G: 'Deployment of artificial neural network for short-term forecasting of evapotranspiration using public weather forecast restricted messages', Agricultural Water Management. 2016; **163**:363–379

[32] Falamarzi Y, Palizdan N, Feng Huang Y, Lee T S: Estimating evapotranspiration from temperature and wind speed data using artificial and wavelet neural network (WNNs), Agricultural Water Management. 2014; **140**:26–36

[33] Yassina M A, Alazbaa A A, Mattar M A: Artificial neural networks versus gene expression programming forestimating reference evapotranspiration in arid climate, Agricultural Water Management. 2016; **163**:110–124

Chapter 3

Sensitivity of Evapotranspiration Models to Onsite and Offsite Meteorological Data for a Ponderosa Pine Forest

Wonsook Ha, Abraham E. Springer,
Frances C. O'Donnell and Thomas E. Kolb

Additional information is available at the end of the chapter

http://dx.doi.org/10.5772/intechopen.68435

Abstract

Evapotranspiration (ET) is a major component of the water budget in most forests, in many cases exceeding 70% of annual precipitation. Due to limitations in time and resources, input data necessary to model ET are not always available for a study site, but offsite data from meteorological networks may be a suitable substitute. In this study, we evaluated three models for estimating ET, Priestly-Taylor (P-T), Shuttleworth-Wallace (S-W), and Penman-Monteith with dynamic stomatal resistance (P-M-d), in a ponderosa pine (*Pinus ponderosa*) forest in northern Arizona where eddy covariance data exist for comparison. We tested the sensitivity of the models to the use of offsite meteorological data from a weather station and offsite soil moisture data from two snow monitoring sites in the SNOTEL network. Onsite data are required for accurate ET estimation with the P-M-d model because of its complexity. Acceptable accuracy in ET estimation required onsite net radiation data for the P-T model and onsite vapor pressure deficit data for the S-W model; other input data can be obtained from nearby offsite weather stations. Errors in ET estimation produced by the use of offsite soil moisture data varied between two nearby SNOTEL sites. Recommendations about the use of offsite data are presented.

Keywords: air temperature, evapotranspiration, net radiation, ponderosa pine forest, vapor pressure deficit

1. Introduction

Evapotranspiration (ET) consists of evaporation from the soil surface and transpiration from the plant canopy [1]. It is one of the major components of the hydrologic cycle, accounting for up to 85% of annual precipitation in some forests [2, 3]. Forest ET affects the frequency and timing of water saturation in soils [4]. Therefore, accurate estimation of ET is needed to predict stream flow emanating from forestland, to investigate hydrological processes, and to manage water resources [4, 5]. Forest ET is known to be hard to quantify [6]. It is a complex hydrological process [7, 8], which is influenced by interactions between the atmosphere, soil, and plant canopy [9, 10].

ET can be directly measured by lysimeters, water balance methods, and eddy covariance systems [6, 11–13]. Direct measurement of ET is difficult and requires large amounts of time, labor, and funding [14, 15]. The eddy covariance method is considered the most reliable, but frequently is deficient from a lack of surface energy balance closure [16]. In the absence of direct measurements, ET can be modeled using climatic and ecosystem data [17–20]. Depending on the model used, estimating ET requires input data such as net radiation, air temperature, vapor pressure deficit, wind speed, and soil moisture. Net radiation is the difference between incoming and outgoing radiation [21], and a key variable for calculating potential ET. Air temperature, wind speed, and vapor pressure deficit control atmospheric conditions that impact ET [22]. Vapor pressure deficit also regulates stomatal resistance, which directly affects ET [23]. Soil moisture controls the availability of water for ET and is especially important in arid and semi-arid ecosystems, where ET is primarily water limited [24, 25]. The use of ET equations with fewer input variables is recommended when complete climatological data cannot be obtained [26]. Meteorological models that use onsite input data can produce estimates of ET similar to measurements by eddy covariance and are less expensive [27].

Installing equipment to measure onsite climatological and soil moisture data may still require more resources than are available to forest and water managers. Publicly available climatological data from weather stations may provide a reasonable substitute. Not all weather stations include the full set of measurements necessary to model ET. The U.S. Surface Climate Reference Network provides data from 114 stations in the USA that record all necessary meteorological data. Other networks, such as Ameriflux (ameriflux.ornl.gov), provide additional coverage. Soil moisture data needed to convert potential ET to actual ET are more limited, distributed across a range of government and academic networks, and summarized in the Texas A&M University North American Soil Moisture Database (http://soilmoisture.tamu.edu/).

In the absence of a nearby weather station with a complete suite of measurements, certain inputs, such as net radiation and vapor pressure deficit, can be calculated using empirical models based on basic weather data [26, 28]. For example, Irmak et al. [28] developed equations to calculate net radiation from input variables such as minimum and maximum air temperatures, measured or predicted solar radiation, inverse relative distance from earth to sun, and mean relative humidity. Tabari et al. [26] derived regression equations to estimate potential ET from air temperature and solar radiation.

The impact on ET estimation of using offsite meteorological data rather than onsite data is unknown for most forest regions. In this study, we estimated ET for a ponderosa pine (*Pinus ponderosa*) forest from three meteorological models (Priestly-Taylor (P-T), Shuttleworth-Wallace (S-W), and Penman-Monteith with dynamic stomatal resistance (P-M-d)) using onsite and offsite data for net radiation, air temperature, vapor pressure deficit, wind speed, and soil moisture content (θ or SMC in this chapter), and compared these estimates to measurements of ET at the same site using the eddy covariance approach as a standard [3, 27]. The objectives of this study are to (1) compare the sensitivity of the P-T, S-W, and P-M-d meteorological ET models to the use of offsite meteorological data, (2) determine if the models can produce reliable results with offsite data, and (3) identify the input data that would result in the largest improvement in ET estimates if measured onsite. Results from the study are valuable to land and resource managers to better understand and predict the forest hydrological cycle using commonly available meteorological data and to prioritize the installation of monitoring equipment with limited resources.

2. Material and methods

2.1. Study site

The study site is a ponderosa pine forest located near Flagstaff, Arizona (Northern Arizona University Centennial Forest: 35°5′20.5″ N, 111°45′43.33″ W, elevation 2180 m a.s.l.). Previously, this site was one of the three sites used to measure carbon and water fluxes in ponderosa pine forests of northern Arizona with the eddy covariance (EC) approach [3]. Thinning, harvesting, and fire did not occur at this site over the last century. The control site had an average leaf area index of 2.3 m^2 m^{-2}, basal area of 30 m^2 ha^{-1}, and tree density of 853 trees ha^{-1} [3, 27, 29]. Location of study site is shown in **Figure 1**.

2.2. ET model selection

We selected the three best performing ET models from Ha et al. [27] based on the model performance statistics of root mean square error (RMSE) and coefficient of determination (R^2) in comparisons with ET measured by the eddy covariance approach (**Table 1**). These three models were the Penman-Monteith dynamic (P-M-d), Priestley-Taylor (P-T), and Shuttleworth-Wallace (S-W) models. The P-M-d model updates the Penman-Monteith ET model by modeling canopy resistance as a function of environmental variables [30]. The P-T model is a simplification of the Penman-Monteith model that only requires inputs of net radiation, air temperature, and the Priestly-Taylor coefficient [31]. The S-W model was developed to estimate ET in sparse canopies and calculates soil evaporation and plant canopy transpiration separately [32]. Each of these models calculates potential ET, which we adjusted to actual ET using a relationship with soil moisture [27, 33]. ET calculations were made at the monthly scale because monthly data were available from weather stations and generally provided more accurate ET estimates for the study site than daily data when compared with ET measured with eddy covariance [27]. Detailed explanation of each model and its use in simulating ET at our study site can be found in Ref. [27].

2.3. Offsite weather station data

Daily offsite meteorological data were obtained from the Western Regional Climate Center (WRCC) available at http://www.wrcc.dri.edu/. Weather data were collected from a weather

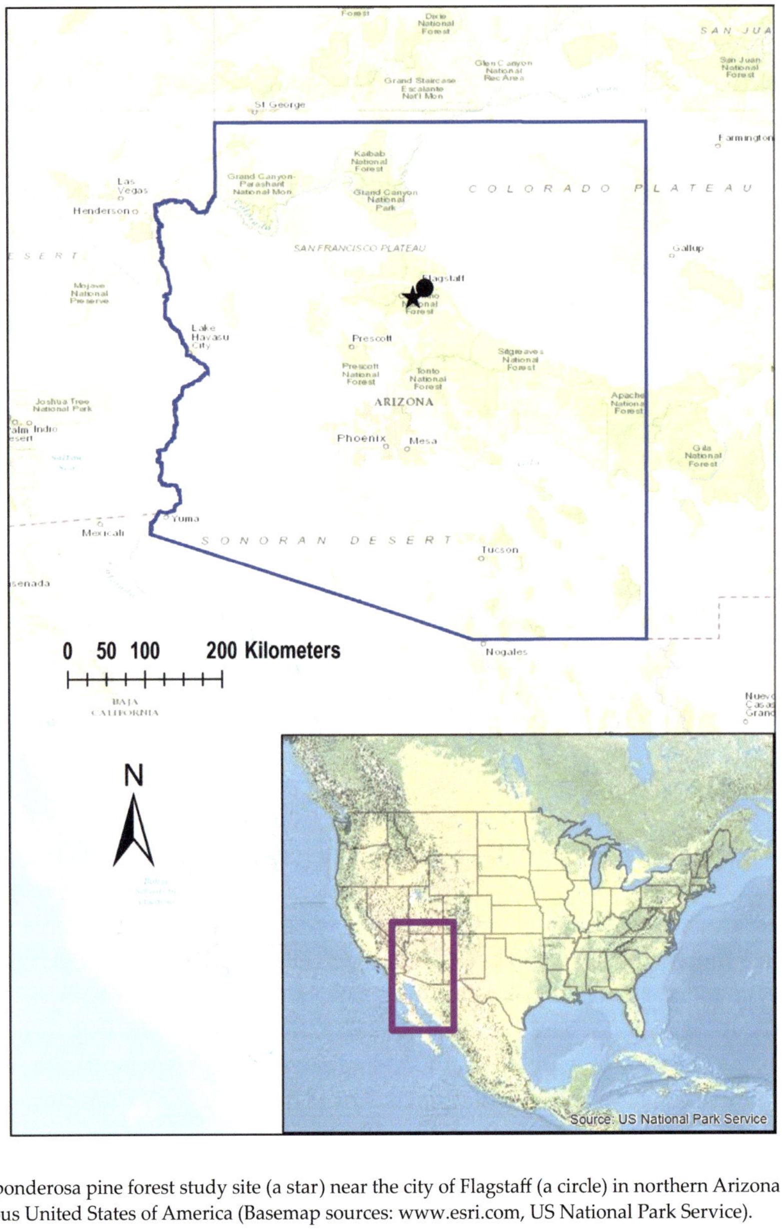

Figure 1. A ponderosa pine forest study site (a star) near the city of Flagstaff (a circle) in northern Arizona with an inset of coterminous United States of America (Basemap sources: www.esri.com, US National Park Service).

station near the Flagstaff Pulliam Airport (Flagstaff 4 SW, AZ; site number: 023009; latitude: 35°8′ N, longitude: 111°40′ W, ground elevation: 2135 m; distance to the study site: 12.2 km). Because net radiation (Rn) data were not available from this station, they were calculated using an empirical equation from the FAO's Irrigation and Drainage paper 56 compiled by Ref. [34]. Inputs were daily offsite maximum and minimum absolute air temperature, actual vapor pressure, dew point temperature, relative humidity, station elevation, solar declination,

C site	RMSE (mm month^{-1})			R^2		
Data	P-M-d	P-T	S-W	P-M-d	P-T	S-W
EC site data only	18.50	**13.19**	15.49	0.81	**0.84**	0.67
Weather station data (Rn, ta, vpd, and u) and site data (SWC, G, α, LAI, etc.)	**20.12**	20.41	21.70	0.80	**0.82**	0.67
Station Rn	18.68	20.23	**15.31**	0.81	**0.82**	0.67
Station ta	19.21	**13.30**	15.47	0.83	**0.84**	0.67
Station vpd	20.15	**13.19**	22.03	0.79	**0.84**	0.67
Station u	19.11	**13.19**	15.49	0.81	**0.84**	0.67
Weather station data and SNOTEL happy jack SMC data (alpha updated)	42.04	**20.08**	30.17	0.26	0.49	**0.62**
Weather station data and SNOTEL Mormon mtn summit SMC (alpha updated)	23.11	51.90	**21.91**	**0.70**	0.50	0.62

Table 1. Comparisons of root mean square error (RMSE) and coefficient of determination (R^2) among three ET models. Numbers in bold indicate the best model for each dataset.

sunset hour angle, inverse relative distance from Earth to sun, sunset hour angle, latitude, and day of the year. Vapor pressure deficit (kPa) was calculated using dew point temperature and relative humidity (%) measured at the station.

Soil moisture data were obtained from the SNOTEL data collection network operated by the Natural Resources Conservation Service (NRCS) of the U.S. Department of Agriculture. Snowpack and climatic data (air and soil temperature and precipitation for all locations and soil moisture data for selected locations) are collected at sites across the Western U.S. Two SNOTEL sites were used, Happy Jack (site number: 969; latitude: 34°45′ N, longitude: 111°25′ W, elevation: 2326 m; distance to study site: 60 km) and Mormon Mountain Summit (site number: 1125; latitude: 34°58′ N, longitude: 111°31′ W, elevation: 2591 m; distance to study site: 32 km). These SNOTEL sites were selected because they were the closest to our study site and they had data available for our period of study (2007–2010). The Mormon Mountain Summit data are available only from June 2008.

2.4. Analysis of error propagation

The error introduced in meteorological models of ET by using offsite station data likely can be reduced if some variables are measured onsite. To aid managers in prioritizing the installation of monitoring equipment with limited resources, an analysis was performed to determine how error in different input data combines to overall model error.

Each model was run at the monthly scale over 4 years (2007–2010) with all onsite input data to establish a baseline result. Then, model runs were performed using offsite data for each input variable and onsite data for the remaining input variables. The percent difference between the single offsite input model and the baseline was calculated. This procedure was repeated with each of the major offsite input variables (net radiation: Rn, air temperature: ta, wind speed: u, vapor pressure deficit: vpd, and soil moisture) to evaluate the error introduced by the use of offsite data.

For each model, the three variables to which the model was most sensitive were selected. Error was introduced to each of the three selected variables in increments of 1% to a maximum of 15% for variables positively related to ET or percent decrease for variables negatively related to ET. Thus, compounding errors acted in the same direction providing a worst case scenario estimate of overall model error. The models were run over ranges of percent error for the variables to determine all combinations of percent error in the three variables that produced 15% model error when averaged overall 4 years of simulation. These results show how variation in the accuracy of input variables affects model results.

3. Results

3.1. Sensitivity of annual ET to offsite data

Figure 2 shows the sensitivity of annual ET predicted by each model to the use of all offsite input data. When compared to eddy ET measurements at the annual scale, the S-W model performed well, but was sensitive to the use of offsite vpd data. The P-M-d model consistently underestimated eddy ET regardless of data sources. The P-T model consistently overestimated ET, and was sensitive to the use of offsite Rn. The use of offsite soil moisture data generally led to underestimation of ET for all models, especially for data from the Happy Jack site. When compared to modeled ET using all onsite data inputs, the use of offsite data inputs only led to significantly increased error in certain cases. For example, accuracy of the S-W and P-T models was reduced by the use of offsite vpd and Rn data, respectively.

3.2. Sensitivity of monthly ET to offsite data

ET simulated by the P-T model was sensitive to the use of offsite weather data used to calculate Rn (**Figure 3(a)**). The difference in ET predicted by the P-T model using all offsite data except for Rn was greater than a 15% threshold of acceptability, especially during summer (data not shown). In contrast, using offsite Rn data produced acceptable predictions of ET by the P-M-d and S-W models (**Figure 3(a)**).

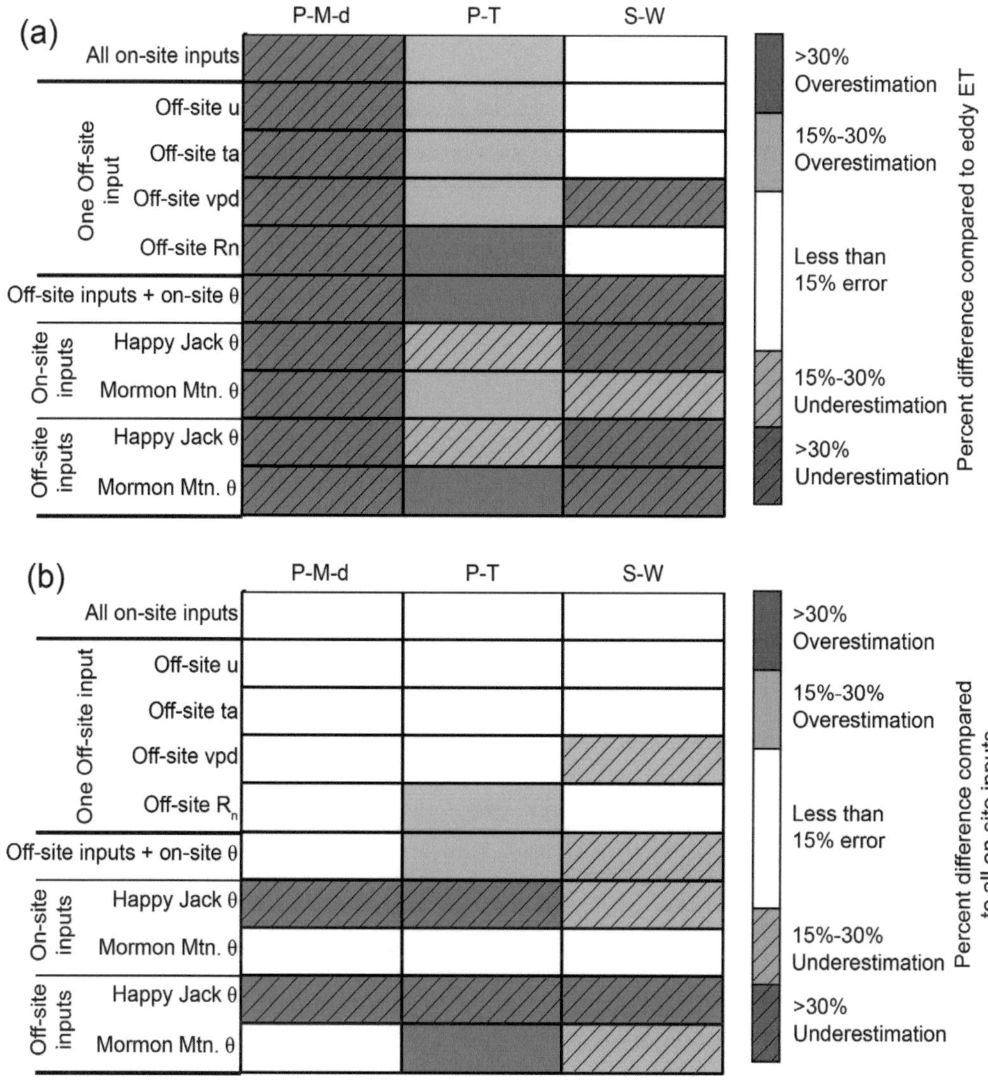

Figure 2. Effect of source of meteorological data for variables, wind speed (u), air temperature (ta), vapor pressure deficit (vpd), and net radiation (Rn), and soil moisture content (θ), on modeled ET as compared to (a) eddy covariance ET measurements and (b) modeled ET with all onsite inputs. Percent difference is calculated at the annual scale for three models, Penman-Monteith with dynamic canopy resistance (P-M-d), Priestly-Taylor (P-T), and Shuttleworth-Wallace (S-W), with all onsite meteorological inputs, a single offsite meteorological input, all offsite meteorological inputs, and offsite soil moisture from two SNOTEL sites, Happy Jack and Mormon Mountain Summit.

The P-M-d and P-T models were sensitive to the use of offsite ta data, whereas the S-W was not (**Figure 3(b)**). Inaccuracy in predicted ET from the P-M-d model was largest during winter when total ET is close to zero.

The P-M-d and S-W models were highly sensitive to the use of offsite vpd data, whereas the P-T model was not because it does not use vpd as an input parameter (**Figure 3(c)**). Inaccuracy

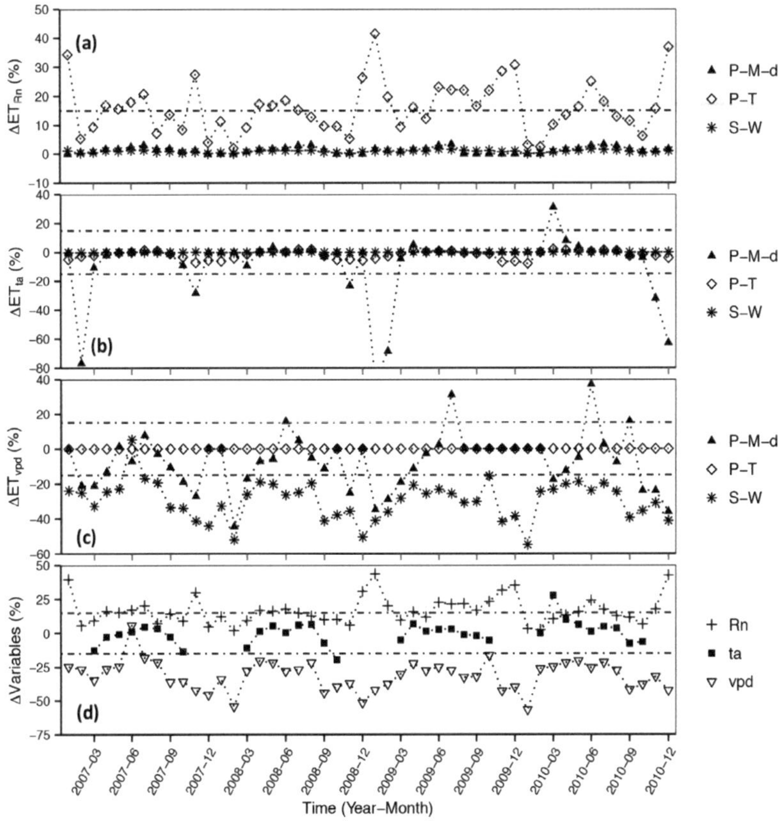

Figure 3. Percent difference in ET from models using all offsite input data and substitution of each input variable with onsite data (subscripted; Rn: net radiation, ta: air temperature, vpd: vapor pressure deficit) for the Penman-Monteith with dynamic canopy resistance (P-M-d), Priestly-Taylor (P-T), and Shuttleworth-Wallace (S-W) models (a–c); and percent difference in seasonal variation between offsite and SNOTEL sites Rn, ta, vpd (d). Dashed horizontal lines indicate ±15% difference lines.

caused by the use of offsite vpd data was greater for the S-W model than the P-M-d model. Sensitivity to the use of offsite u was low for all models (data not shown).

Modeled Rn based on offsite weather data was consistently higher than based onsite Rn (**Figure 3(d)**), with the highest percent difference in summer. The percent difference between offsite and onsite vpd peaked during summer (**Figure 3(d)**). As opposed to Rn, offsite vpd was always lower than onsite vpd (**Figure 3(d)**). Offsite ta data generally provided a good substitute for onsite data, differing by less than 2.5°C throughout the study period. During winter, when temperatures were close to 0°C, the percent difference between offsite and onsite temperatures could be quite large despite small absolute differences. Thus, these values are excluded from **Figure 3(d)**.

The P-T and S-W models were most sensitive to the source of soil moisture content (SMC) data (**Figure 4(a)** and **(c)**). The P-T model was more sensitive than the S-W model, because SMC is

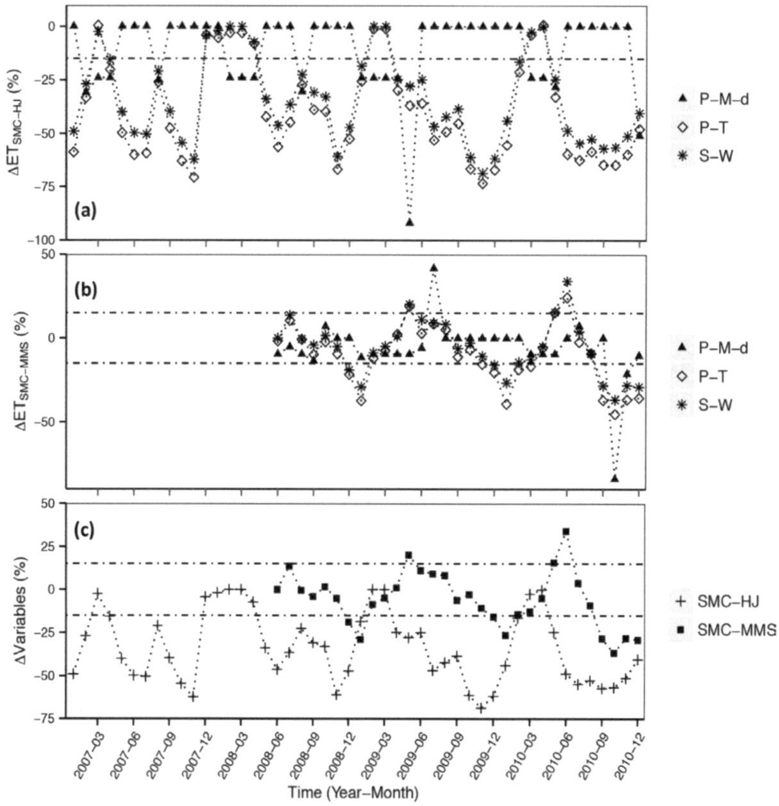

Figure 4. Percent difference in ET from models using all offsite input data and substitution of each input variable with onsite data (subscripted; SMC-HJ: soil moisture content from Happy Jack SNOTEL site, SMC-MMS: soil moisture content from Mormon Mountain Summit SNOTEL site) for the Penman-Monteith with dynamic canopy resistance (P-M-d), Priestly-Taylor (P-T), and Shuttleworth-Wallace (S-W) models (a and b); and percent difference in seasonal variations of soil moisture contents (SMC) between two SNOTEL sites (c). Dashed horizontal lines indicate ±15% difference lines.

included in the calculation of the scaling coefficient α in the P-T model. The Happy Jack (HJ) SMC data produced a larger difference in **ΔET** than Mormon Mountain Summit (MMS) SMC data, which resulted from consistently lower SMC at the HJ site than onsite (**Figure 4(c)**). The SMC at the MMS site was higher than onsite SMC during summer, but lower during winter. SMC data from the MMS site were unavailable between January 2007 and May 2008, so no percent change was calculated during this time (**Figure 4(b)** and **(c)**).

The difference between ET modeled with offsite and onsite input data with the P-M-d model led to overprediction of ET during summer (**Figure 5**) due to the error introduced by vpd (**Figure 3(c)**), and underestimated ET during winter due both to error from vpd and ta (**Figure 3(b)**). The P-T model had a bigger difference than P-M-d between offsite and onsite modeled ET during summer due to error introduced by offsite Rn (**Figure 3(a)**). ET stimulated by the S-W model using offsite data was always lower than using onsite data except for June 2007 (**Figure 5**), a pattern driven by error introduced by offsite vpd.

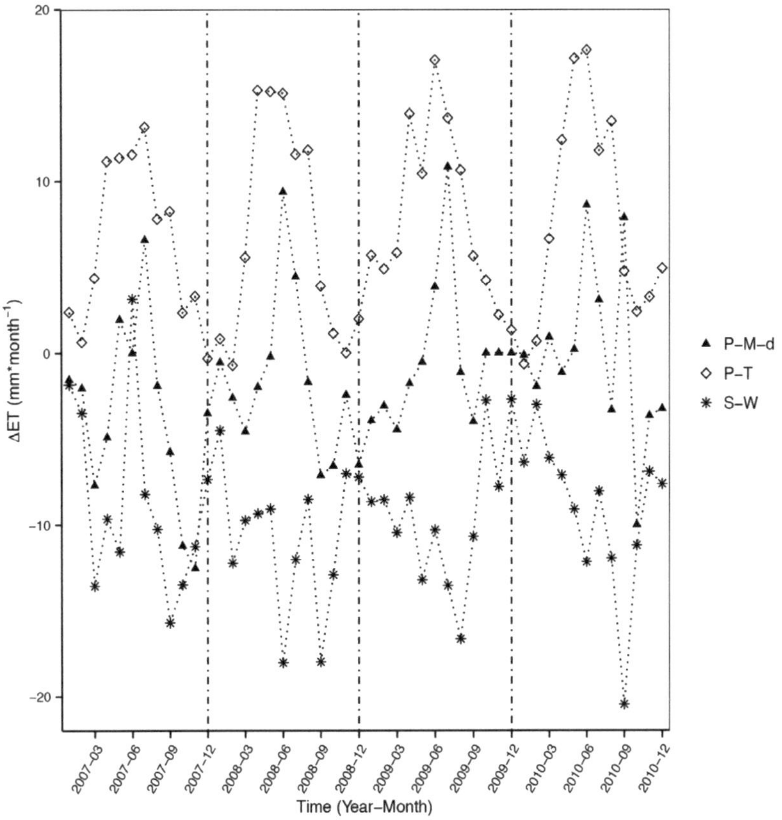

Figure 5. Difference in ET between modeled using offsite weather station input data and modeled with onsite data for the Penman-Monteith with dynamic canopy resistance (P-M-d), Priestly-Taylor (P-T), and Shuttleworth-Wallace (S-W) models.

3.3. Sensitivity of annual ET to input errors

Error in prediction of annual ET by the P-T model was most strongly influenced by error in Rn and, to a lesser extent, SMC (**Figure 6(a)**). Less than 15% error in overall modeled ET was possible with large errors in ta (>15%) and soil SMC (>20%) if onsite Rn (i.e., 0% error in net radiation) is used. Error in the S-W model was most strongly influenced by error in vpd (**Figure 6(b)**). Large errors in soil moisture and temperature were acceptable if vpd is accurate. Because of the threshold responses and complex internal dynamics of the P-M-d model, we conclude that it is not a good choice for this type of error analysis. Therefore, we only considered the S-W and P-T models in this analysis.

3.4. Relationships between onsite and offsite weather data

A strong linear relationship occurred between onsite data and offsite weather- or SNOTEL-station data with R^2 0.72 or higher for all variables, except for u where the R^2 was 0.59 (**Figure 7**).

http://dx.doi.org/10.5772/intechopen.68435

(a)

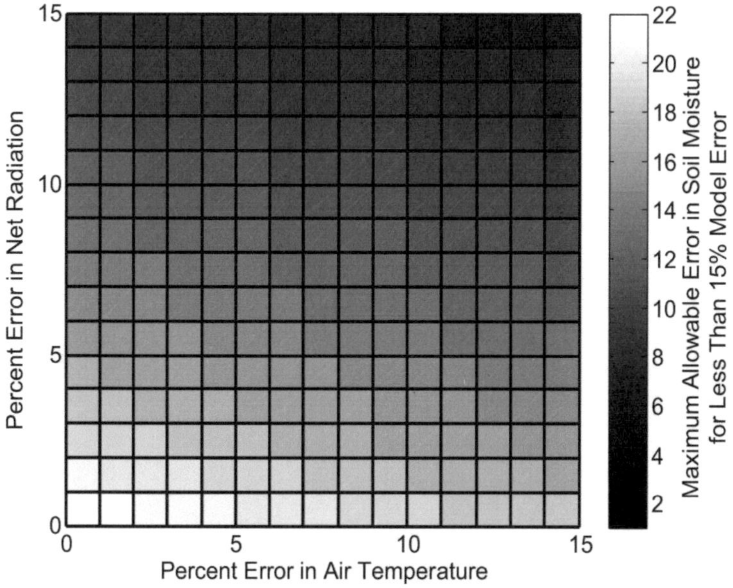

(b)

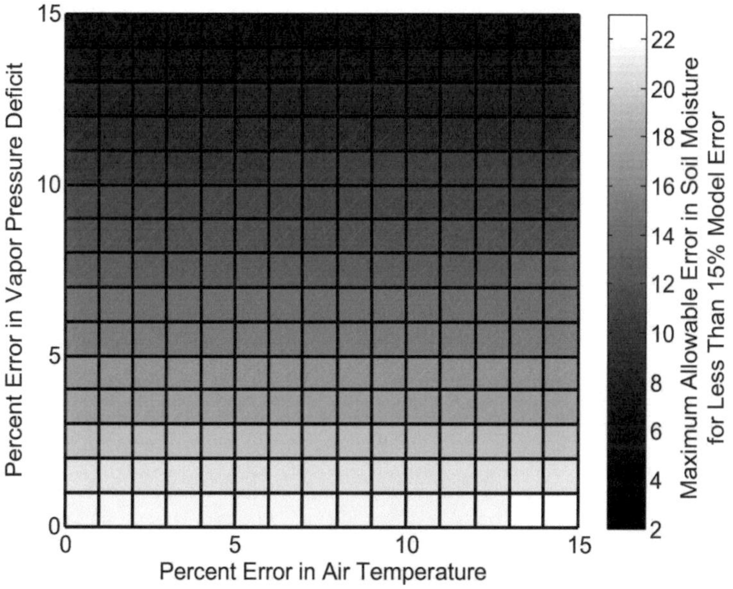

Figure 6. Combination of error in net radiation, air temperature, and soil moisture content that will produce 15% error in the Priestly-Taylor model (a), and vapor pressure deficit, air temperature, and soil moisture that will produce 15% error in the Shuttleworth-Wallace model (b). Model error is calculated at the annual scale over four years (2007–2010). Other model inputs were measured onsite.

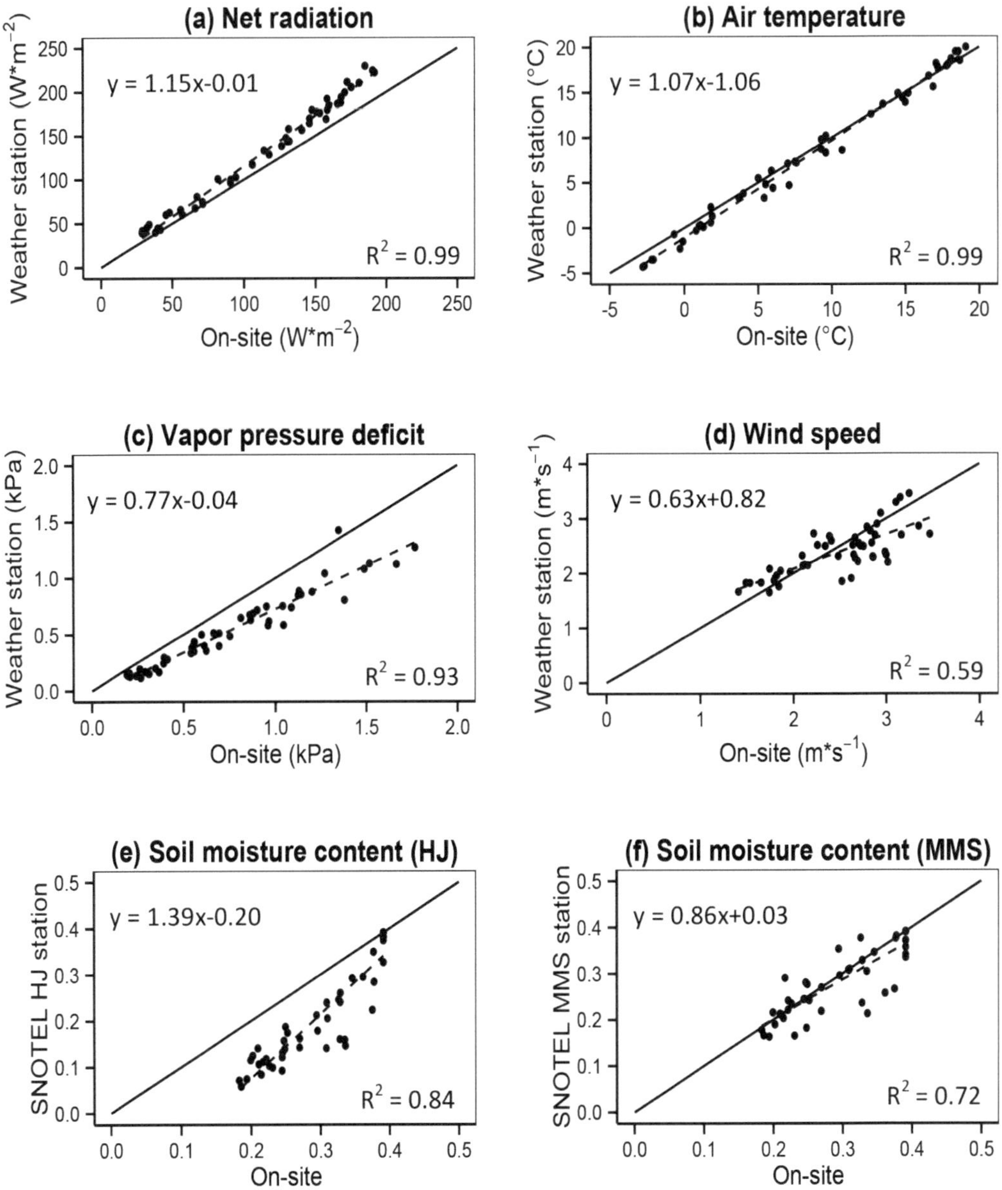

Figure 7. 1:1 relationship between onsite eddy tower versus offsite weather station input variables for (a) net radiation (W m^{-2}), (b) air temperature (°C), (c) vapor pressure deficit (kPa), and (d) wind speed (m s^{-1}). Soil moisture content data from eddy tower versus Happy Jack (HJ) station (e) and Mormon Mountain Summit (MMS) (f) are also shown. The solid line indicates a 1:1 line and the dashed line is a regression line.

4. Discussion

The performance of three meteorological ET models (P-M-d, P-T, and S-W) in predicting ET measured by eddy covariance for the ponderosa pine forest in our study was affected by the source of input data. The P-M-d model performed well with offsite Rn data. In most cases, the P-T model performed well when offsite soil moisture content (SMC) data obtained from local SNOTEL sites replaced onsite SMC data. Likewise, the S-W model performed well when onsite Rn, ta, and u data were replaced by offsite data from a nearby weather station. A previous study [27] reported that when using all onsite data, the ET predictions from the S-W model were the closest to ET measured by eddy covariance among five meteorological models, likely because our study site contains a mixture of surface layers (pine canopy, grass, bare soil), which the S-W model was designed to simulate [35–37].

Sensitivity to offsite meteorological data varied among models. The P-T model was sensitive to the use of offsite weather data used to predict Rn (**Figure 2(a)**). The P-M-d model was sensitive to the use of offsite ta and vpd data (**Figure 2(b)** and **(c)**). The S-W model was sensitive to the use of offsite vpd data (**Figure 2(c)**). From this result, we conclude that accurate measurement of vpd is important to properly estimate ET using S-W and P-M-d models. Models using Rn predicted from offsite weather data calculated consistently higher ET than models using onsite Rn data (**Figure 2(d)**), resulting in overestimation of ET by the P-T model, especially in summer (**Figure 2(a)**), when ET is highest and is the most important seasonal component to annual ET.

The suitability of offsite data as inputs into meteorological ET models depended on the variable and model. Offsite weather station data provided suitable estimates of onsite air temperature (**Figure 2(d)**) and therefore, air temperature was not a major source of error in all models for this study site. The sensitivity analysis (**Figure 5**) illustrates that errors in air temperature up to 15% are not likely to lead to large errors in ET estimates. Overestimation of winter ET occurs in the P-M-d model when onsite ta is used (**Figure 2(b)**), but winter ET is a small component of total annual ET. The dynamic model of stomatal conductance in the P-M-d model goes to zero when air temperature is below zero, and the overestimation by this model is the result of a threshold response when onsite air temperature is below zero and offsite air temperature is not. The vpd was consistently lower in the offsite data than the onsite data leading to overestimation of ET by the S-W model without a seasonal pattern. The effect of vpd on the P-M-d model is more complex because increases in vpd increase ET by increasing evaporative demand and decrease ET by reducing stomatal conductance. Thus, overestimation or underestimation of ET is possible when onsite vpd data are replaced with offsite data.

5. Conclusions

We investigated sensitivity to mixtures of on and offsite data inputs of three widely used models for calculating ET (P-M-d, P-T, and S-W) for a ponderosa pine forest where ET was measured previously by eddy covariance. The complexity of the P-M-d model makes it highly prone to inaccurate ET predictions with offsite data. The P-T and S-W models can provide reliable estimates of ET with selected input variables measured onsite and combined with offsite data for other inputs. Because measurement of some of these input data is expensive and difficult at specific field sites, resources should be devoted to onsite measurement of variables that model are most sensitive to. We found that offsite data from a nearby weather station were sufficient for air temperature and wind speed, whereas model accuracy in predicting ET was improved by onsite measurements of Rn for the P-T model, and vpd for the S-W model. The feasibility of using offsite soil moisture data depends on the proximity and similarity of the offsite monitoring location to the study site. Although, we did not determine the sensitivity of the models to data sources in forest conditions beyond one unmanaged ponderosa pine stand, our methods can be applied to other situations, where baseline measurements of ET by eddy covariance or other approaches exist.

Acknowledgements

Authors acknowledge the financial support from the WaterSMART Applied Science Grants for the Desert Landscape Conservation Cooperative of Bureau of Reclamation (Grant number: R12AC80912). Authors also thank anonymous reviewers for their proofs to improve this book chapter.

Author details

Wonsook Ha[1]*, Abraham E. Springer[2], Frances C. O'Donnell[3] and Thomas E. Kolb[4,5]

*Address all correspondence to: wonsook-ha@uiowa.edu

1 Department of Civil and Environmental Engineering, University of Iowa, Iowa City, IA, United States

2 School of Earth Sciences and Environmental Sustainability, Northern Arizona University, Flagstaff, AZ, United States

3 Department of Civil Engineering, Auburn University, Auburn, AL, United States

4 School of Forestry, Northern Arizona University, Flagstaff, AZ, United States

5 Merriam-Powell Center for Environmental Research, Northern Arizona University, Flagstaff, AZ, United States

References

[1] Liang X, Lettenmaier DP, Wood EF, Burges SJ. A simple hydrologically based model of land surface water and energy fluxes for general circulation models. Journal of Geophysical Research: Atmospheres. 1994;**99**:14415-14428.

[2] Arain MA, Black TA, Barr AG, Griffis TJ, Morgenstern K, Nesic Z. Year-round observations of the energy and water vapour fluxes above a boreal black spruce forest. Hydrological Processes. 2003;**17**:3581-3600.

[3] Dore S, Montes-Helu M, Hart SC, Hungate BA, Koch GW, Moon JB, et al. Recovery of ponderosa pine ecosystem carbon and water fluxes from thinning and stand-replacing fire. Global Change Biology. 2012;**18**:3171-3185.

[4] Vose JM, Sun G, Ford CR, Bredemeier M, Otsuki K, Wei X, et al. Forest ecohydrological research in the 21st century: What are the critical needs?. Ecohydrology. 2011;**4**:146-158.

[5] Ruhoff AL, Paz AR, Aragao LEOC, Mu Q, Malhi Y, Collischonn W, et al. Assessment of the modis global evapotranspiration algorithm using eddy covariance measurements and hydrological modelling in the Rio Grande basin. Hydrological Sciences Journal. 2013;**58**:1658-1676.

[6] Sun G, Alstad K, Chen J, Chen S, Ford CR, Lin G, et al. A general predictive model for estimating monthly ecosystem evapotranspiration. Ecohydrology. 2011;**4**:245-255.

[7] Kumar M, Raghuwanshi NS, Singh R, Wallender WW, Pruitt WO. Estimating evapotranspiration using artificial neural network. Journal of Irrigation and Drainage Engineering. 2002;**128**:224-233.

[8] Traore S, Wang Y-M, Kerh T. Artificial neural network for modeling reference evapotranspiration complex process in Sudano-Sahelian zone. Agricultural Water Management. 2010;**97**:707-714.

[9] Kondo J, Saigusa N, Sato T. A parameterization of evaporation from bare soil surfaces. Journal of Applied Meteorology. 1990;**29**:385-389.

[10] Reynolds JF, Kemp PR, Tenhunen JD. Effects of long-term rainfall variability on evapotranspiration and soil water distribution in the Chihuahuan desert: A modeling analysis. Plant Ecology. 2000;**150**:145-159.

[11] Rana G, Katerji N. Measurement and estimation of actual evapotranspiration in the field under mediterranean climate: A review. European Journal of Agronomy. 2000;**13**:125-153.

[12] Law BE, Falge E, Gu L, Baldocchi DD, Bakwin P, Berbigier P, et al. Environmental controls over carbon dioxide and water vapor exchange of terrestrial vegetation. Agricultural and Forest Meteorology. 2002;**113**:97-120.

[13] Vaughan PJ, Trout TJ, Ayars JE. A processing method for weighing lysimeter data and comparison to micrometeorological ETo predictions. Agricultural Water Management. 2007;**88**:141-146.

[14] Irmak S, Howell TA, Allen RG, Payero JO, Martin DL. Standardized ASCE Penman-Monteith: Impact of sum-of-hourly vs. 24-hour timestep computations at reference weather station sites. Transactions of the ASAE. 2005;**48**:1063-1077.

[15] Temesgen B, Eching S, Frame K. Comparing net radiation estimation methods: CIMIS versus Penman–Monteith. Journal of Irrigation and Drainage Engineering. 2007;**133**:265-271.

[16] Twine TE, Kustas WP, Norman JM, Cook DR, Houser PR, Meyers TP, et al. Correcting eddy-covariance flux underestimates over a grassland. Agricultural and Forest Meteorology. 2000;**103**:279-300.

[17] Monteith JL. Evaporation and the environment. Symposium of the Society of Exploratory Biology. 1965;**19**:205-234.

[18] Wilson KB, Hanson PJ, Mulholland PJ, Baldocchi DD, Wullschleger SD. A comparison of methods for determining forest evapotranspiration and its components: Sap-flow, soil water budget, eddy covariance and catchment water balance. Agricultural and Forest Meteorology. 2001;**106**:153-168.

[19] Allen RG, Tasumi M, Trezza R. Satellite-based energy balance for mapping evapotranspiration with internalized calibration (metric)—model. Journal of Irrigation and Drainage Engineering. 2007;**133**:380-394.

[20] Baldocchi D. Measuring fluxes of trace gases and energy between ecosystems and the atmosphere—The state and future of the eddy covariance method. Global Change Biology. 2014;**20**:3600-3609.

[21] Su Z, Roebeling RA, Schulz J, Holleman I, Levizzani V, Timmermans WJ, et al. Observation of hydrological processes using remote sensing. In: Wilderer P, editor. Treatise on water science. Oxford, UK: Academic Press; 2011. pp. 351-399.

[22] Morillas L, Villagarcía L, Domingo F, Nieto H, Uclés O, García M. Environmental factors affecting the accuracy of surface fluxes from a two-source model in mediterranean drylands: Upscaling instantaneous to daytime estimates. Agricultural and Forest Meteorology. 2014;**189–190**:140-158.

[23] Dickinson RE, Henderson-Sellers A, Rosenzweig C, Sellers PJ. Evapotranspiration models with canopy resistance for use in climate models, a review. Agricultural and Forest Meteorology. 1991;**54**:373-388.

[24] Williams CA, Albertson JD. Contrasting short- and long-timescale effects of vegetation dynamics on water and carbon fluxes in water-limited ecosystems. Water Resources Research. 2005;**41**:W06005. DOI: doi:10.1029/2004WR003750

[25] Chen X, Rubin Y, Ma S, Baldocchi D. Observations and stochastic modeling of soil moisture control on evapotranspiration in a Californian Oak Savanna. Water Resources Research. 2008;**44**:W08409. DOI: doi:10.1029/2007WR006646

[26] Tabari H, Grismer M, Trajkovic S. Comparative analysis of 31 reference evapotranspiration methods under humid conditions. Irrigation Science. 2013;**31**:107-117.

[27] Ha W, Kolb TE, Springer AE, Dore S, O'Donnell FC, Martinez Morales R, et al. Evapotranspiration comparisons between eddy covariance measurements and meteorological and remote-sensing-based models in disturbed ponderosa pine forests. Ecohydrology. 2015;**8**:1335-1350.

[28] Irmak S, Irmak A, Jones JW, Howell TA, Jacobs JM, Allen RG, et al. Predicting daily net radiation using minimum climatological data. Journal of Irrigation and Drainage Engineering. 2003;**129**:256-269.

[29] Dore S, Kolb TE, Montes-Helu M, Sullivan BW, Winslow WD, Hart SC, et al. Long-term impact of a stand-replacing fire on ecosystem CO_2 exchange of a ponderosa pine forest. Global Change Biology. 2008;**14**:1801-1820.

[30] Stewart JB. Modeling surface conductance of pine forest. Agricultural and Forest Meteorology. 1988;**43**:19-35.

[31] Priestley CHB, Taylor RJ. On the assessment of surface heat flux and evaporation using large scale parameters. Monthly Weather Review. 1972;**100**:81–92.

[32] Shuttleworth WJ, Wallace JS. Evaporation from sparse crops—an energy combination theory. Quarterly Journal of the Royal Meteorological Society. 1985;**111**:839-855.

[33] Flint AL, Childs SW. Use of the Priestley-Taylor evaporation equation for soil water limited conditions in a small forest clearcut. Agricultural and Forest Meteorology. 1991;**56**:247-260.

[34] Allen RG, Pereira LS, Raes D, Smith M. Crop evapotranspiration (guidelines for computing crop water requirements). Rome, Italy: FAO; 1998. 333 p.

[35] Ortega-Farias S, Poblete-Echeverría C, Brisson N. Parameterization of a two-layer model for estimating vineyard evapotranspiration using meteorological measurements. Agricultural and Forest Meteorology. 2010;**150**:276-286.

[36] Lagos L, Martin D, Verma S, Irmak S, Irmak A, Eisenhauer D, et al. Surface energy balance model of transpiration from variable canopy cover and evaporation from residue-covered or bare soil systems: Model evaluation. Irrigation Science. 2013;**31**:135-150.

[37] Zhu G, Su Y, Li X, Zhang K, Li C. Estimating actual evapotranspiration from an alpine grassland on Qinghai-Tibetan plateau using a two-source model and parameter uncertainty analysis by Bayesian approach. Journal of Hydrology. 2013;**476**:42-51.

Chapter 4

Evapotranspiration in Northern Agro-Ecosystems: Numerical Simulation and Experimental Comparison

Watcharee Ruairuen, Gilberto J. Fochesatto,
Marco Bittelli, Elena B. Sparrow,
Mingchu Zhang and William Schnabel

Additional information is available at the end of the chapter

http://dx.doi.org/10.5772/intechopen.68347

Abstract

Evapotranspiration and near-surface soil moisture dynamics are key-entangled variables regulating flux at the surface-atmosphere interface. Both are central in improving mass and energy balances in agro ecosystems. However, under the extreme conditions of high-latitude soils and weather pattern variability, the implementation of such coupled liquid and vapor phase numerical simulation remain to be tested. We consider the nonisothermal solution of the vapor flux equation that accounts for the thermally driven water vapor transport and phase changes. Fully coupled flux model outputs are compared and contrasted against field measurements of soil temperature, heat flux, water content, and evaporation in a subarctic agroecosystem in Alaska. Two well-defined hydro-meteorological situations were selected: dry and wet periods. Numerical simulation was forced by time series of incoming global solar radiation and atmospheric surface layer thermodynamic parameters: surface wind speed, ambient temperature, relative humidity, precipitation, and soil temperature and soil moisture. In this simulation, soil parameters changing in depth and time are considered as dynamically adjusted boundary conditions for solving the set of coupled differential equations. Results from this evaluation give good correlation of modeled and observed data in net radiation (R_{net}) (R^2 of 0.92, root mean square error (RMSE) of 45 W m^{-2}), latent heat (0.70, RMSE of 53 W m^{-2}), and sensible heat ($R^2 = 0.63$, RMSE = 32 W m^{-2}) during the dry period. On the other hand, a poor agreement was obtained in the radiative fluxes and turbulent fluxes during the wet period due to the lack of representation in the radiation field and differences in soil dynamics across the landscape.

Keywords: coupled fluxes, evapotranspiration, soil dynamics, vapor transport, numerical simulation

1. Introduction

Northern latitudes have been identified as a region where global climate change will have earlier and stronger impacts than in other regions of the world [1–4]. Most of the region is underlain by discontinuous permafrost or perennially frozen ground in which temperatures remain below 0°C for at least two consecutive years. An active layer on top of the permafrost experiences seasonal thaws and is the primary dominant subsurface component of the land-atmosphere system [5]. Under climate warming scenario, much of this terrain would be vulnerable to subsidence, particularly in ice-rich areas of relatively warm, discontinuous permafrost, and shrinking ponds and lakes [3, 6–10]. All these changes will potentially alter the exchange of surface energy, water, and carbon cycles in high-latitude ecosystems [11] and consequently, the response at regional level to the atmosphere system.

Soil moisture plays a critical role in the surface energy balance and water cycle in these regions [1, 12, 13]. It is widely recognized that the soil moisture confined in a thin layer underneath the land surface influences the partitioning of the surface energy fluxes simultaneously modifying surface thermal conductance and rates of evaporation [14]. An example of such an important role is the control of precipitation transfer into the soil and the partitioning of incoming solar radiation into latent, sensible, and ground heat fluxes [15, 16]. In addition, soil moisture and temperature status affect biological processes such as soil microbial activity, seed germination, and plant growth. These variables in turn also affect water and nutrition absorption and solute transport in soil. In a climate change scenario, high-latitudes soils will experience increased summer dryness as climate warming progresses, changing therefore atmospheric vapor pressure conditions and thereby enhancing evapotranspiration (ET) rate. In terms of seasonal effects, inadequate snowmelt infiltration or rainfall during spring and early summer often causes crop water stress and reduction in yield of small grains [17, 18] in agricultural activities of subarctic regions. Therefore, understanding the variation of evapotranspiration (ET) and its impact on crop growth becomes of absolute importance because it mainly controls the available soil water and, therefore, is a limiting factor in agriculture productivity and sustainability. As a result, continuous monitoring of soil water content and soil temperature is a priority in the fields of agronomy and hydrology [19].

Several modeling studies have focused on soil carbon reservoirs (e.g. [20–22]) and permafrost degradation in natural ecosystems across the circumpolar region [21, 23]; nevertheless, agro-ecosystem has not been taken into consideration until a recent study by Ruairuen et al. [24]. Despite the mentioned complexities in the soil medium, similarities between high latitude and mid-latitude agricultural soils exist mainly during the growing season. This allows for making use of models that are currently in use for mid-latitude agricultural settings. In this case, a fully coupled differential equation system considering both soil temperature and vertical soil moisture distribution, and their interactions are utilized to bring emphasis on the sub-medium transport in contrast to most large-scale ecosystem models where one or two soil layers are used to simulate soil moisture dynamics in ecosystem models (e.g. [25]).

In this study, we use the numerical model developed by Bittelli et al. [26], which fully couples heat and water transport to deduce the coupled water and heat transport across the soil medium forced by radiation and meteorological conditions. As demonstrated in mid-latitudes studies by Bittelli et al. [26], this approach enables numerically stable, energy and mass conservation equation solution in terms of the external forcing and boundary conditions. Such an approach requires a modeling framework that incorporates the interactions among meteorological variables (e.g., air temperature, relative humidity, precipitation, and solar radiation) and soil properties (e.g., soil temperature, soil moisture, and soil water potential) into the coupled numerical model. We also conducted a field experiment to measure net radiation, sensible heat flux, soil moisture, and soil temperature, and use those measured parameters to calculate evapotranspiration in order to compare with simulated results.

2. Materials and methods

2.1. Field experiment

The field study was conducted at the Fairbanks Experiment Farm (FEF) of the University of Alaska Fairbanks (UAF) Agricultural and Forestry Experiment Station (AFES) Fairbanks Alaska (**Figure 1**), USA (64°51′16.6″ N, 147°51′36.4″ W, 150 m above sea level) during summer 2013. The soil within the lysimeter plots, established in a previous study, was used for this study because large amounts of data (i.e., soil moisture, soil temperature, and soil moisture

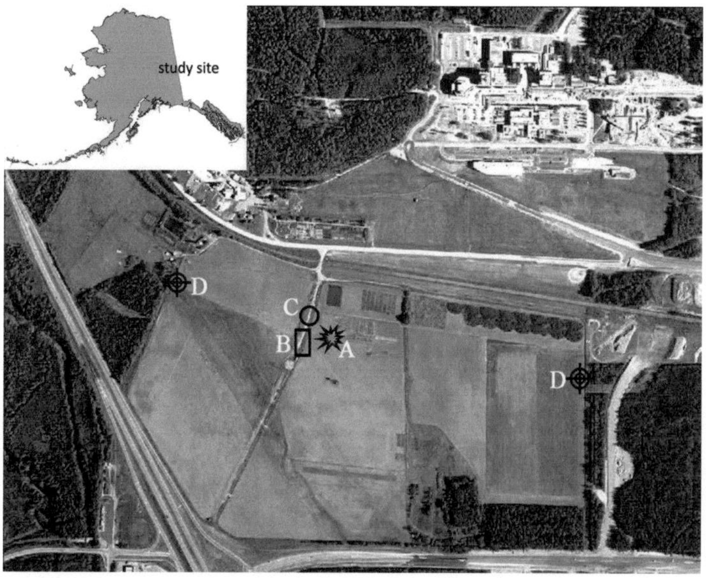

Figure 1. Geographic location of the Northern Latitude Study Site (panel left on the top) and Fairbanks Experiment Farm (FEF) at the UAF-AFES within the University of Alaska Fairbanks Campus (panel top right). Bottom panel is the detail on the location of the instrumentation in the study site. The farm dimensions are more than 1 km on East to West direction and about 600 m North to South. Eddy-Covariance (EC) tower (A), lysimeter plot (B), meteorological station (C), large aperture scintillometer (LAS) Scintec BLS900 (D). Airborne survey photo was provided by the UAF Department of Design and Construction obtained by AeroMap Inc. summer of 2003.

potential) were available. The soil composition was sandy loam with 66 sand, 29 silt, and 5% clay, and with the available water holding capacity of about 0.18–0.36 $m^3 m^{-3}$ that was determined from a soil moisture characteristic curve [24]. Parameters for soil hydraulic properties to be introduced in the numerical simulation are listed in **Table 1**. Volumetric soil moisture content was measured *in situ* using three soil moisture sensor (10HS; Decagon Devices Inc., Pullman, WA, USA) at 5, 10, and 20 cm. The sensor has 14.5 cm long prongs, 3.3 cm wide, so the 5-cm sensor spanned the depth 3.3–6.7 cm, the 10-cm sensor spanned the depth 8.3–11.3 cm, and the 20-cm sensor spanned the depth 18.3–21.3 cm. Gravimetric samples were also taken to calibrate the 10HS sensor [24]. Soil moisture was continuously measured with an interval of 30 min. The bulk density was measured in the tested site (**Table 1**). The soil temperature (S-TMB-M006, Onset HOBO Corporation, Bourne, MA, USA) was measured at depths of 5, 10, and 20 cm below the soil surface and used to obtain the ground heat flux in this study.

The observation-based meteorological parameters included air temperature (Tair), relative humidity (RH), air pressure, wind speed (u), and direction, and precipitation at 2 m height above the ground were obtained at 1-min intervals at the experimental station. One-min recordings of these data were averaged to obtain hourly data for input to the simulation.

An independent measure of evapotranspiration was determined using Penman-Monteith method and the more continuous series of data available on this period. Sensible heat flux was measured locally (i.e., ecosystem scale) by means of and eddy-covariance (EC) instrument and processed considering signal distortions under all weather conditions [27] and, at landscape scale, based on a large aperture scintillometer (LAS) [28–30].

2.2. Model implementation

2.2.1. Model description (PSP_coupled)

The numerical model was coded in Python and is set in a time-evolving one-dimensional simulation of coupled flow of liquid water, heat, and water vapor. The model description can be found in Bittelli et al. [26, 31]. **Figure 2** shows the models' conceptual scheme indicating the coupled layers and driving boundary conditions, i.e., soil temperature, liquid water

Soil property	Value
Bulk density (g cm^{-3})	0.7
Air entry potential (J kg^{-1})	−1.5
Mass sand (kg kg^{-1})	0.66
Mass silt (kg kg^{-1})	0.29
Mass clay (kg kg^{-1})	0.05
Saturated moisture content; θ_s ($m^3 m^{-3}$)	0.56
b value (−)	−3.1
K_s (kg s m^{-3})	7.2×10^{-4}

Table 1. Soil properties within lysimeter.

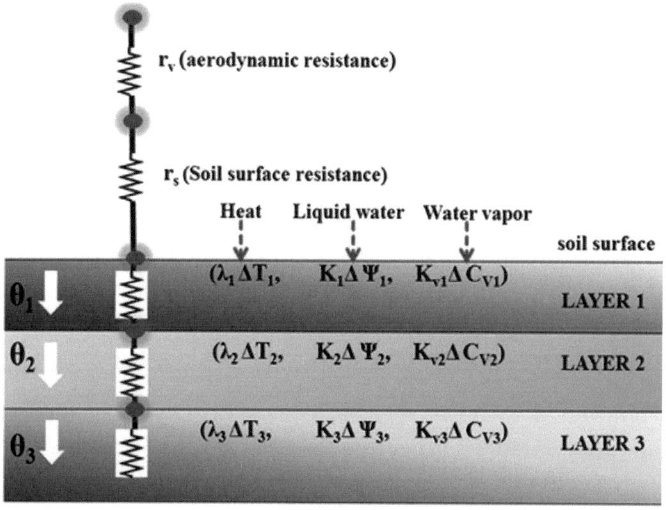

Figure 2. Scheme of the computational grid with the driving force terms (temperature, soil water potential, and soil vapor concentration), the soil conductivities and the resistances involved at the soil-atmosphere interface (adapted from Bittelli et al. [26]). The black dot is the mass balance for heat flow and water flux at a given node.

volumetric concentration, and soil resistive and conductive terms. The model computes the soil energy budget. The PSP-coupled model is composed of ten modules with two input data files and one main program file. One data file contains the soil data, and the other one contains the weather data. Model modules include main.py, PSP_boundary.py, PSP_public.py, PSP_soil.py, PSP_coupled1D.py, PSP_readDataFile.py, PSP_longWaveRadiation.py, PSP_grid.py, PSP_plot.py, PSP_plotEnergy.py, PSP_ThomasAlgoritmh.py, soil.txt, and Weather.txt.

The main file is main.py, which contains the calls to other embedded subroutines listed. The module PSP_boundary defines the initial and boundary conditions. The PSP_public contains all variables that are read by all modules such as latitude, longitude, altitude, albedo, atmospheric pressure and clay content, initial soil temperature, and soil matric potential. The PSP_soil is written to define the soil properties. The PSP_couple1D is the module that implements the solver for the different flux equations. The PSP_longWaveRadiation is for computing the long wave radiation component of the radiation balance at the soil surface. The PSP_grid module is for building the computational grid and PSP_ThomasAlgorithm for solving the system of equations. The PSP_plot and PSP_plotEnergy are modules for visualizing the data input and output from the model.

2.2.2. Initial setting for model simulation

The initial conditions for dry and wet periods were calculated using *in situ* data. To implement the simulation scenarios, two data files need to be created "soil.txt" and "weather.dat." The soil.txt file is required for data input of soil properties such as soil depth (the model set up from 0 to 1.5 m), the saturated soil moisture, residual water content, soil hydraulic properties, and soil matric potential (**Table 1**). The soil file can be used to simulate the two periods; however, additional settings indicating initial conditions should be modified in the PSP_public module. The weather file is required at hourly weather parameters such as solar radiation, air temperature, precipitation, relative humidity, and wind speed as an input to calculate related parameters. Time step is set to 300 s and input data of 1-h resolution.

The mass of sand, silt, clay, and bulk density was obtained from *in situ* measurements. The parameter for the hydraulic properties was obtained from Cambell and Shiozawa [32], K_s is saturated hydraulic conductivity, θ_s is the saturated soil moisture content, and b is constant value.

The PSP_public is the file that needs to be adapted in all parameters that are read by all modules for the given area in which the simulation is carried out. In this case, the FEF site-specific information was input including latitude, longitude, and altitude. Moreover, the soil initial conditions such as soil water potential, soil temperature, and albedo for dry and wet scenarios needed to be applied into this module (**Table 2**).

In this study, the value of albedo was set as 0.2 for the dry period [33], while a value of 0.15 was applied for the wet period in agriculture land in subarctic region according to previous studies in the same agricultural setting [34].

Parameters	Dry period	Wet period
Number of days	9	6
Soil temperature (°C)	17.2	15.0
Soil water potential (J kg^{-1})	−30	−6
Albedo	0.20	0.15
Initial soil moisture (m^3 m^{-3})	0.28	0.30

Table 2. Initial setting for model simulation.

3. Simulation scenarios

As mentioned previously, the model was applied to two selected periods (dry and wet) in an agricultural land described in Section 2 during the summer 2013, and its performance was evaluated. In addition and based on local meteorological information, it was verified that the atmospheric boundary layer developed forced by surface and near surface flow conditions without presence of multilayered thermal inversions [35]. However, this condition is difficult to maintain when precipitation arises. The dry period (no precipitation event) spun from 26 July to August 3 (Julian day 207–216) and wet period from 25 to 30 August (Julian day 237–242).

3.1. Dry period

The experimental data were taken from the plots that monitored soil moisture and soil temperature. The meteorological parameters measured about 10 m away from the plot for the dry period are given in **Table 3** and **Figure 3a**. The hourly average vapor pressure deficit (VPD) was 1.18 kPa. Mean-hourly global radiation was 215 W m^{-2}. The average air temperature was 21°C with maximum of 30.7°C and minimum of 9.8°C also reported in this period. The average RH was approximately 58% and wind speed of 2.36 m s^{-1}.

Parameters	Dry	Wet
Mean T_{air} (°C)	20.66 ± 5.32	11.78 ± 3.41
Mean T_s (°C)	18.96 ± 3.2	16.64 ± 1.51
T_s at the beginning period (°C)	18.3	15.2
Mean RH (%)	57.74 ± 17.26	76.72 ± 16.76
Mean VPD (kPa)	1.18 ± 0.76	0.36 ± 0.32
Mean u (m s^{-1})	2.36 ± 1.07	1.68 ± 0.82
Mean solar radiation (W m^{-2})	215.23 ± 221.69	131.31 ± 172.42
Mean soil moisture (m^3 m^{-3})	0.2098 ± 0.0015[a]	0.2097 ± 0.0035[a]
Precipitation (mm)	–	37.60

Soil temperate at 5 cm was considered in the study.
[a]Soil moisture from 5 cm depth.

Table 3. Main meteorological conditions, soil properties, and surface characteristics during periods under study.

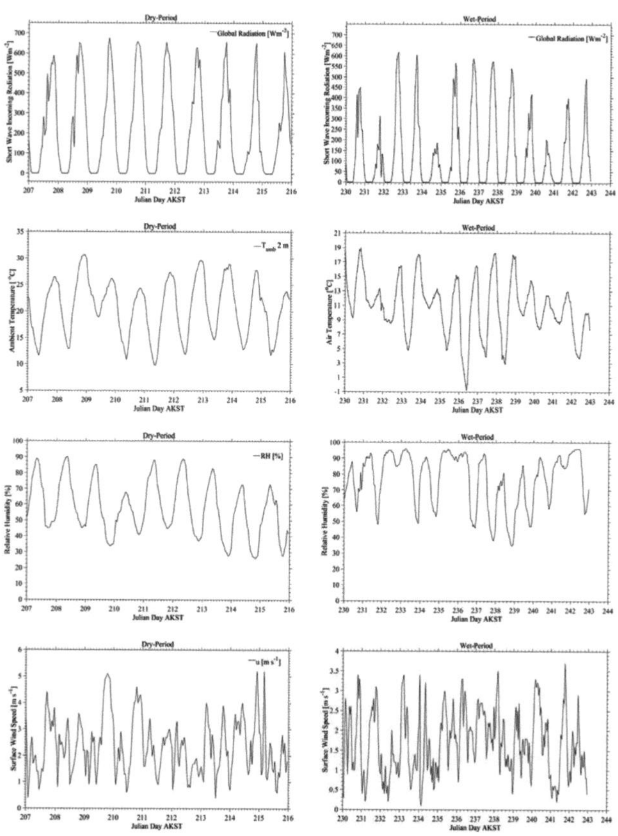

Figure 3. Global radiation, relative humidity, air temperature, and wind speed measured in the atmospheric surface layer from top to bottom at the experimental plot during the dry period (left panel; 26 July–3 August 2013—Julian day 207–215) and wet period (right panel; 18–30 August 2013—Julian 230–242).

3.2. Wet period

The meteorological conditions during the wet period 18–30 August 2013 (Julian 230–242) was cooler than the dry period in terms of an average hourly air temperature and soil temperature (**Figure 3b**), while there was a slight difference in solar radiation compared to the dry period (see **Table 3**). The RH was approximately 77% with low level of VPD (0.36 kPa) on average during the wet period (**Table 3**). A total precipitation of 37.60 mm was also reported in this period. However, only data during 25–30 August 2013 (Julian 237–242) are used as the wet period for the simulation in this study.

4. Results and discussion

4.1. Net radiation

Net radiation (R_{net}) is the main input in the energy balance equation driving evapotranspiration process. Giving this importance, a comparison between the net radiation observed (R_{net} obs) and modeled (R_{net} mod) was performed. **Figure 4a** shows simulated and measured hourly data of net radiation as a function of time for the dry period (26 July–3 August 2013). Results show a good agreement between modeled and observed R_{net} with the correlation coefficient (R^2) of 0.92 and root mean square error (RMSE) of 45 W m^{-2} (**Figure 5a**) during the dry period. There are about 6 days out of nine total days that the model performed remarkably well to simulate the R_{net} and the maximum difference did not exceeded 157 W m^{-2} on 27 July (Julian day 208).

Overall, R_{net} mod overestimated R_{net} obs by about 16%. Similar results have been reported by Ortega-Farias et al. [36] in a commercial vineyard where the R_{net} simulation shows a good agreement with the measured R_{net} ($R^2 = 0.92$) with the RMSE less than 48 W m^{-2}. On the other hand, a lower correlation between R_{net} obs and R_{net} mod was found during the wet period with the R^2 of 0.72 and RMSE of 67 W m^{-2} (**Figure 5b**). However, there were two days from 25 to 26 August that the model simulated very well compared with the measured R_{net}, while it was underestimated in the following 2 days (27–28 August) and overestimated in the last 2 days during the wet period (**Figure 4b**). The maximum underestimation for the entire time series was about 85 W m^{-2}, and the maximum overestimate was 199 W m^{-2}.

The high and low correlation between measured and modeled R_{net} in different periods can be related to the cloudiness variability in high latitudes. This may be explained based on the variability of incoming solar radiation between those periods as we can see in **Figure 6**. An average of solar radiation of 215 W m^{-2} with the maximum and minimum of 680, 150 W m^{-2} respectively were reported in the dry period, while a lower mean and maximum R_{net} were found in the wet period (**Table 3**). Ortega-Farias et al. [37] indicated that errors in the calculation of R_{net} over a well-irrigated festuca grass were associated with the estimation of atmospheric radiation under cloudy sky conditions. Therefore, to summarize this point, R_{net} mod was able to estimate net radiation with a good degree of precision during the dry period than during the wet period.

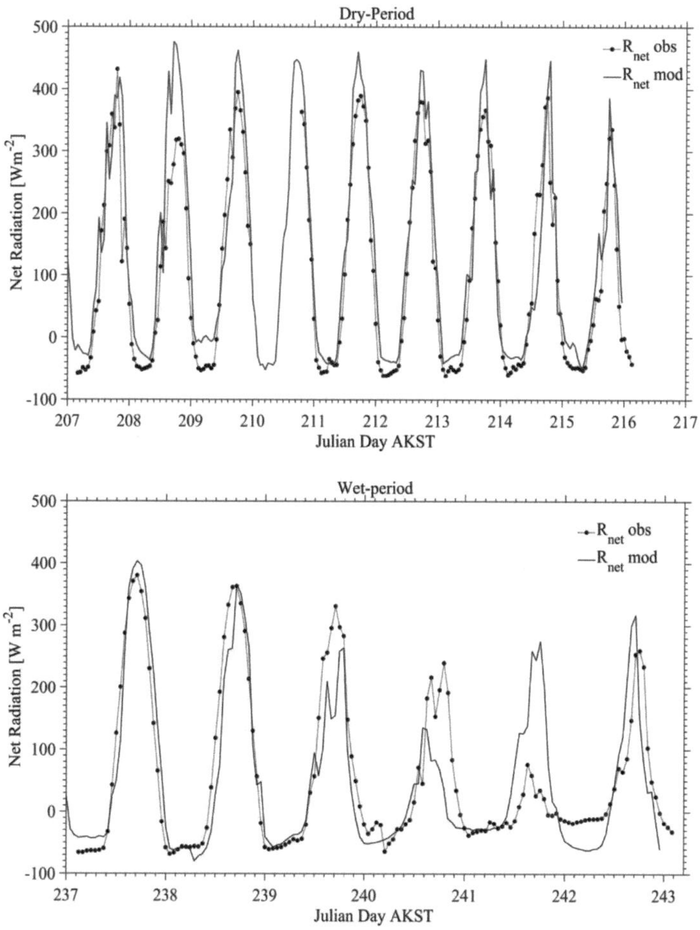

Figure 4. Simulated and measured hourly data of net radiation during the dry and wet period as a function of time. (Top panel) Dry period from 26 July to 3 August 2013 (Julian day 207–216) (bottom panel) wet period from 25 to 30 August 2013 (Julian day 237–242).

4.2. Latent heat

Figure 7 shows simulated and observed latent heat flux (LE) as a function of time during dry and wet periods. High rates of solar radiation heating the soil surface caused the soil to lose water vapor to evaporation from the surface. During nighttime, there was negative conduction to cool the soil surface, and LE became negative due to condensation. The LE was better predicted by the numerical model during the dry period, with an R^2 of 0.70 and RMSE of 53 W m^{-2} than the wet period (R^2 = 63 and RMSE = 58 W m^{-2}). There were 2 days that the model overestimated LE, but the time series followed each other, while 1 day obtained similar values and 3 days of modeled LE did not correlate with the observed LE. The maximum difference between observed and simulated LE was about 200 W m^{-2}. Cumulative ET from the observed quantities reached 22 mm with an average of 2.44 mm d^{-1}, while cumulative ET

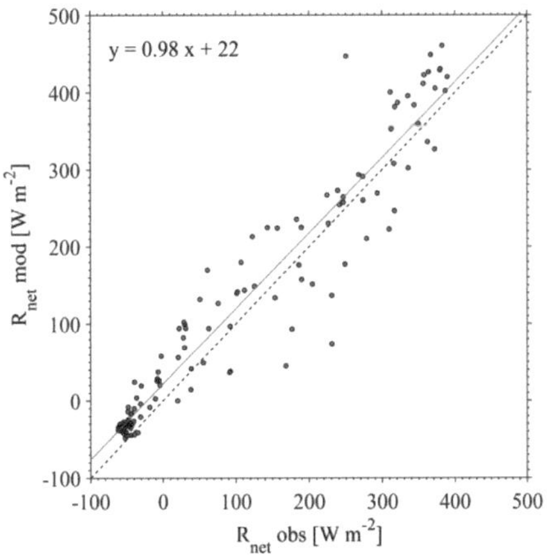

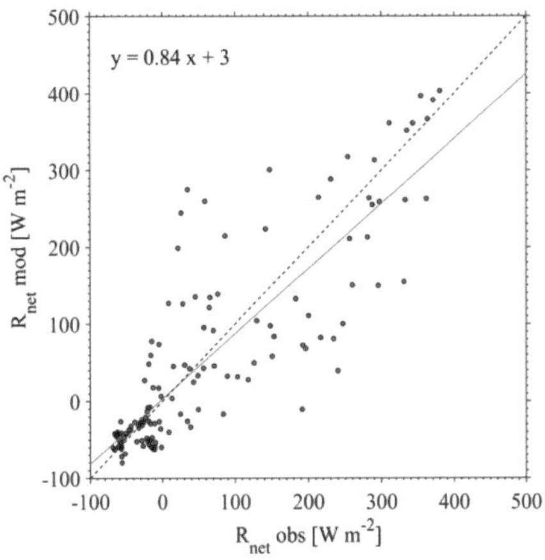

Figure 5. Comparison between the hourly observation of net radiation (R_{net} obs) and the simulation of net radiation (R_{net} mod) during the dry and wet period. (Top panel) Dry period from 31 July to 3 August 2013 (Julian day 210–215) and during the wet period (bottom panel) from 25 to 30 August 2013 (Julian day 237–242).

from the vsimulation was 15 mm with the mean of 1.67 mm d^{-1} for the dry period. In contrast, during the wet period, the cumulative ET from the observed was about 6 mm over 6 days, whereas only about 2 mm was found from the simulation of ET cycles. An average precipitation observed was 1.0 mm d^{-1}, while only 0.3 mm d^{-1} was found for the simulation during the wet period.

4.3. Ground heat fluxes

The average daytime ground heat flux (*G*) contribution to vapor flux was in the order of 26 W m^{-2} (ranging from 1 to 79 W m^{-2}), while the value output from the simulation was on average 25 W m^{-2} (ranging from 0.8 to 49 W m^{-2}; **Figure 8**). In **Figure 8a**, modeled *G* resulted as overestimated for the first two days, and then closely approached the *G* observed at day 3, and underestimated the observations during the last 2 days of the dry period. The maximum reported difference was

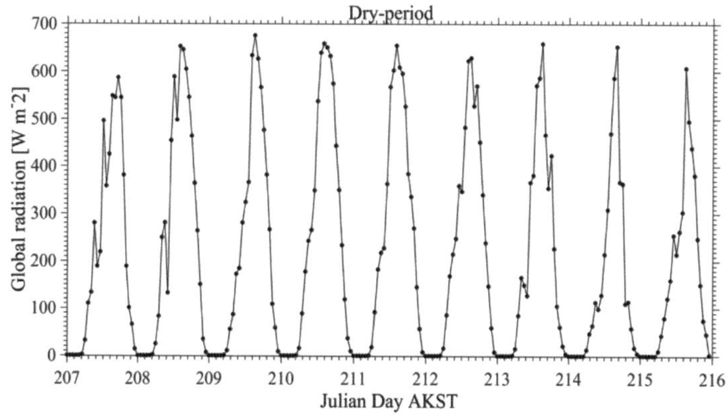

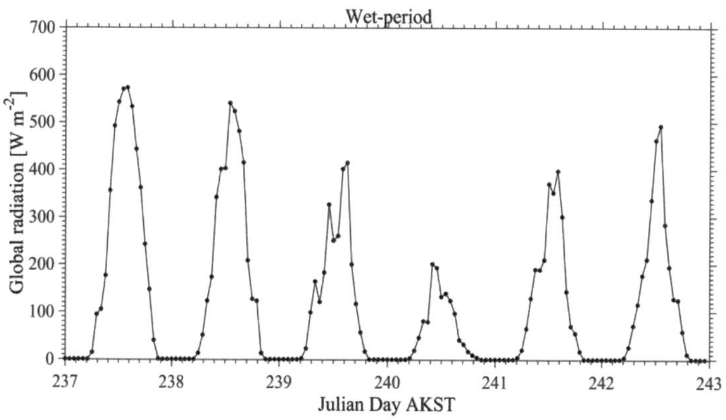

Figure 6. Solar radiation as a function of time during dry and wet periods. (Top panel) Dry period from 26 July to 3 August 2013 (Julian day 207–215) and (bottom panel) wet period from 25 to 30 August 2013 (Julian day 237–242).

44 W m^{-2} during the overestimating period, whereas the same values were 32 W m^{-2} found during the underestimating period. On the contrary, the simulation of G during the wet period was overestimated for the entire period (**Figure 8b**) with the maximum difference of more than 100 W m^{-2} during daytime.

4.4. Sensible heat

The sensible heat (H) fluxes obtained from measurements and from the simulation for agricultural field were compared during the dry and wet period. The time series composite of hourly diurnal variation during dry and wet was illustrated in **Figure 9**. In the dry period, results showed a good correlation between H measured by EC and simulated from models with $R^2 = 0.63$ and RMSE = 32 W m^{-2}, whereas a lower correlation was found in H observed by LAS (R^2 of 0.52 and RMSE of 40 W m^{-2}) (**Figure 9a**). Nevertheless, LAS measurements represent a larger fluxing footprint area than EC-derived fluxes, and the simulations output converge well between these two scale-dependent observations. Overall simulated outputs overestimated H from EC except

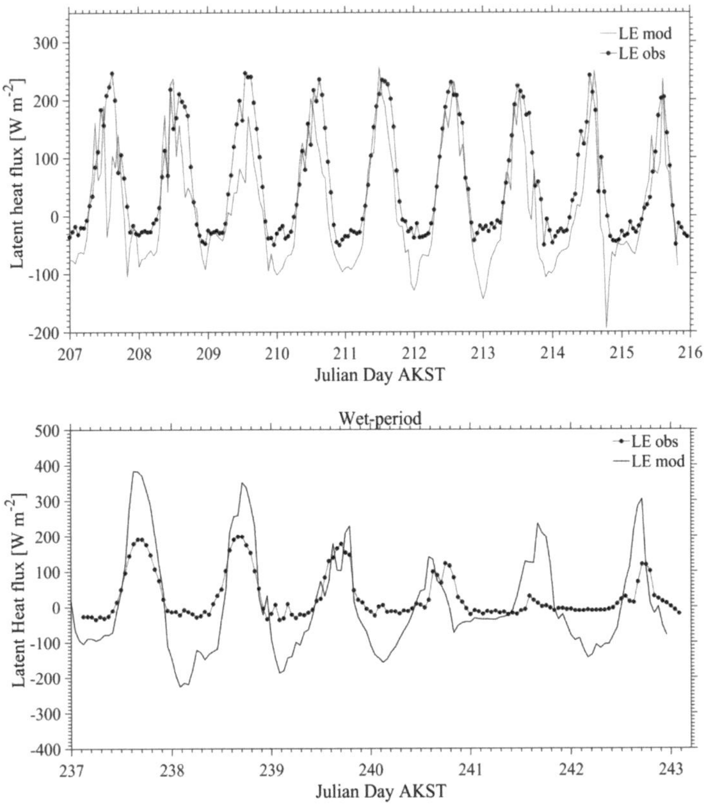

Figure 7. Evapotranspiration time series observed and modeled. (Top panel) Dry period from 26 July to 3 August 2013 (Julian day 207–216) (bottom panel) wet period from 25 to 30 August 2013 (Julian day 237–242).

in the morning where they were closer in value. The ratio of *H* by EC and simulated was about 0.47, while ratio of *H* modeled versus *H* by LAS was 0.82. The maximum of *H* from model, EC, and LAS were 139, 110 and 169 W m^{-2}, respectively, during midday.

Concerning the wet period, very poor correlation was found between *H* measured by LAS and simulated outputs for *H* ($R^2 = 0.15$, RMSE =13 W m^{-2}). It should be noted that *H* by EC was not illustrated during this period because of insufficient data for the analysis. The time-series composite of hourly diurnal variation of *H* from both methods are illustrated in **Figure 9b**. Mean hourly *H* observed was 30 W m^{-2}, while only 15 W m^{-2} was reported from modeled *H*. In general, *H* showed higher values than *H* simulated by about 51% with the maximum difference of 36 W m^{-2} during the wet period.

4.5. Soil temperature and soil moisture

Soil temperature was measured at the same depth as soil moisture. **Figure 10** shows simulated and observed soil temperature as a function of time during dry and wet periods. During the dry period, the value of soil temperature from experiment reached 17°C and was higher than

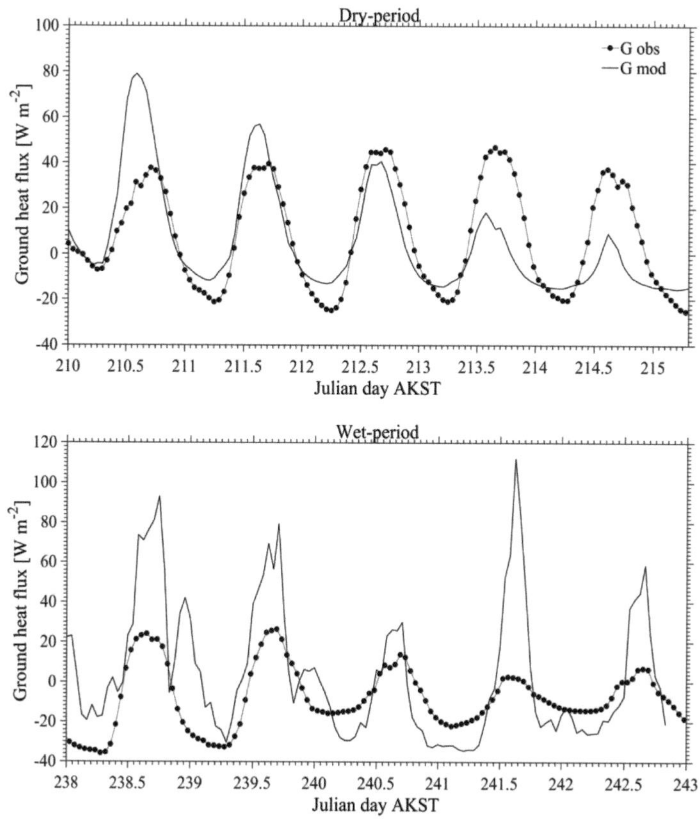

Figure 8. Time series of simulated and estimated ground heat flux. (Top panel) Dry period from 29 July to 2 August 2013 (Julian day 210–214) (bottom panel) wet period from 26 to 30 August 2013 (Julian day 238–242).

the simulated 11°C for the same depth at the beginning of the simulation period. However, after midday in the first day, the simulated value was higher than observed soil temperature over entire dry period (**Figure 10a**). The soil temperature was better predicted by the numerical model during the wet period, with an R^2 of 0.59 and RMSE of 1.82°C than the dry period ($R^2 = 0.47$ and RMSE = 3.27°C). The maximum difference between observed and simulated soil temperature was about 8°C during mid of the day in dry period and the times series followed each other. There was only a day that the model underestimated in a wet period, while most of the time, simulated values overestimated the observed data (**Figure 10b**). A large difference of the soil temperature can cause differences in the land-atmosphere temperature gradient that affect the ground heat flux as described in the previous sections. However, in this case, we found the model in good agreement in predicting soil temperatures for this environment.

The soil moisture was measured in the plot at three depths. From the measurements, we found that the high moisture content was in the lower depth than in the surface layer. The initial soil moisture during the dry period above the soil surface was about 0.28 $m^3\ m^{-3}$ (5 cm depth). The numerical model gave a lower level soil moisture around the same depth with a large difference of 0.21 $m^3\ m^{-3}$ when compared to the observed data. Some disagreement

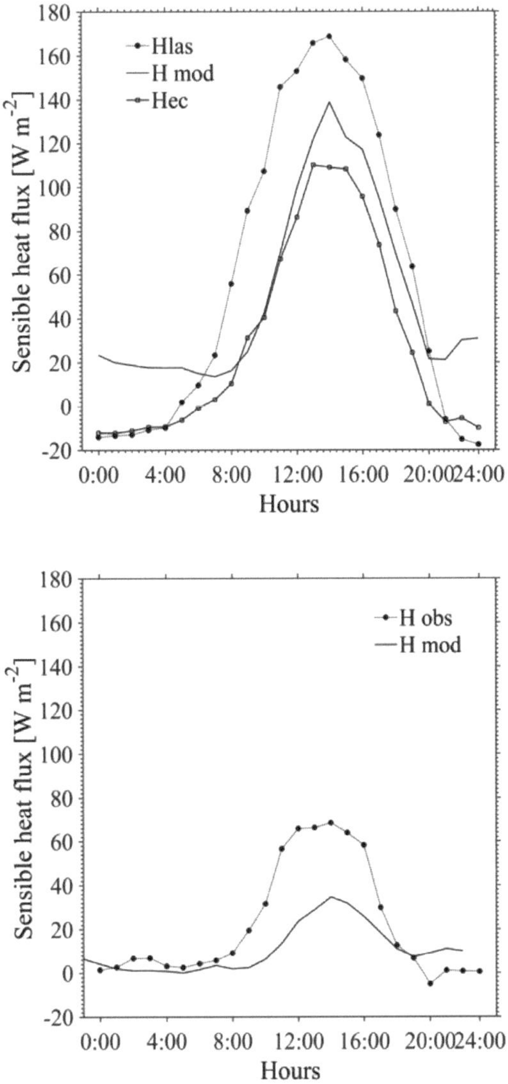

Figure 9. Time-series composite of hourly diurnal variation of sensible heat flux during (top panel) dry and (bottom panel) wet period.

between modeled and observed data was also found during the wet period where the measurement of soil moisture from the plot was 0.30 m^3 m^{-3} for the first day; however, the simulation gave a lower value of soil moisture with a difference larger than 0.13 m^3 m^{-3}. In addition, an underestimation of simulated soil moisture potential is also reported in this study during both periods. This could be due to the lack of representation on the hydraulic properties of the soils especially for the subarctic soil. This is important because the hydraulic conductivity versus the soil moisture potential curve is highly nonlinear and, therefore, the flow of soil moisture from the upper layer to the lower layer in a wet period leads to a large decrease in hydraulic conductivity and liquid water distribution [26, 31], while during the dry period,

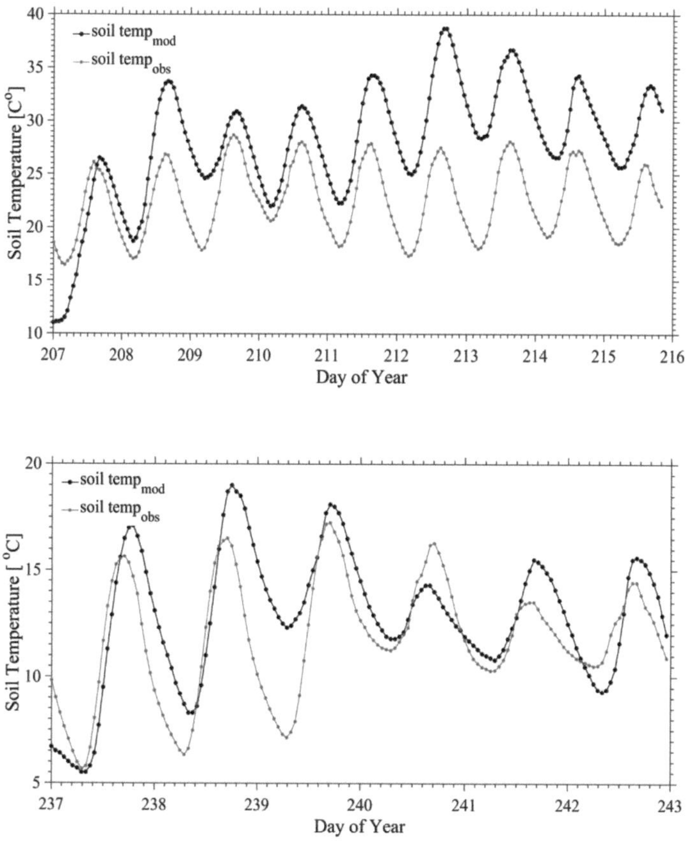

Figure 10. Soil temperature time series observed and modeled. (Top panel) Dry period from 26 July to 3 August 2013 (Julian day 207–216), (bottom panel) wet period from 25 to 30 August 2013 (Julian day 237–242).

the soil moisture was more constant along the depths with less dependence on the liquid fraction. The soil moisture content fluctuated through the day according to the vapor flux as reported previously [38–40]. As such, improvements in the model's representation of both soil moisture and soil moisture potential in order to have an optimal simulation output. There is also a need to further study the vertical soil properties along the soil depth under agricultural land in subarctic region. Evaluation of simulation performance for subarctic soil and weather seems that more parameters might be needed to improve model simulation because of influence of the permafrost, soil properties across landscape and weather variability.

5. Conclusions

An effort was undertaken to simulate fluxes from surface-atmosphere exchanges based on numerically solving the coupled vapor and liquid water differential equations prescribing soil properties and turbulent exchange parameters. Numerical simulation was forced by meteorological data, radiation, and precipitation from a high-latitude agricultural farm. Similarly, dynamic boundary conditions were introduced throughout the simulation including soil temperature, soil moisture, and soil hydraulic properties.

After examining simulation outputs and comparing them to collocated micrometeorological data, it can be concluded that time series of fluxes during the dry period seemed to be reproduced fairly better than those obtained during the wet period. In general, R_{net} has a good agreement between modeled and observed data for both periods with RMSE of 45 W m^{-2} in a dry period and 48 W m^{-2} during a wet period. The LE also was well predicted by the model with RMSE not exceeding 53 W m^{-2}. This difference in turbulent fluxes agrees well with other studies in the same area over highly heterogeneous terrain with further canopy complexity [41, 42]. On the other hand, *G* was overestimated with the maximum difference of more than 100 W m^{-2} in a wet period and 44 W m^{-2} in a dry period. In addition, the measured *H* by EC and LAS instruments correlated well with the model in terms of the RMSE being in the range of 32–40 W m^{-2} which falls within the interval of fluxing difference across landscapes observed on the same area for heterogeneous surfaces [41, 42]. However, only the soil temperature correlated better during a wet period than a dry period, while the soil moisture and soil moisture potential was underestimated compared to the observed values. The low correlation in the wet period was due to significant influences of the synoptic variability introducing large changes in cloud coverage and precipitation that are difficult to reproduce by a single point one-dimensional model formulation. Nevertheless, dry conditions that are by far the most stringent conditions for agriculture sustainability reproduces well.

There are still several parameters such as the presence of vegetation above the soil, the swell, and shrink of soil that need to be investigated more in depth and the most important factor is the hydraulic properties of soil and its variability across landscape. This variable is more complicated, and there are many steps to reach an approximately correct value. In the current study, existing values were applied from previous work done around the same study site, while some other values were obtained from field and laboratory experiments. However, it is known that soils in agro-ecosystems tend to experience large changes in some of these properties and, therefore, are difficult to capture. This factor needs to be taken into account when implementing this model over unnatural setting systems.

Finally, based on current numerical model outputs and field experimental observations allowed identification new challenges in northern agro-ecosystems. Improved representation of soil dynamics is necessary to improve fidelity in the simulations, and also there is a need to establish better strategies to compare single-point numerical modeling to scale-dependent micrometeorological observations. In addition, a large deviation in simulated soil profiles and heat exchanges reveals the highly heterogeneous nature of an aerodynamically simple terrain considered in terms of atmospheric observations.

Acknowledgements

The authors thank the support of the Agricultural and Forestry Experiment Station (AFES) School of Natural Resources and Extension at the University of Alaska Fairbanks and the support of the Department of Agricultural Sciences at the University of Bologna, Italy. Ms. Watcharee Ruairuen was funded by Suratthani Rajabhat University, Thailand. GJ Fochesatto was supported by the Geophysical Institute, University of Alaska Fairbanks. Elena B. Sparrow was supported by the International Arctic Research Center and Alaska EPSCOR, also at the University of Alaska Fairbanks. Mingchu Zhang was supported by USDA NIFA Multi state NC-1179 funding through the Department of Agriculture and Horticulture, School of Natural Resources and Extension, University of Alaska Fairbanks

Author details

Watcharee Ruairuen[1*], Gilberto J. Fochesatto[2], Marco Bittelli[3], Elena B. Sparrow[4], Mingchu Zhang[5] and William Schnabel[6]

*Address all correspondence to: ruairuen@gmail.com

1 Department of Environmental Sciences, Faculty of Science and Technology, Surattani Rajabhat University, Surattani, Thailand

2 Department of Atmospheric Sciences, Geophysical Institute and College of Natural Science and Mathematics, University of Alaska Fairbanks, Fairbanks, Alaska, USA

3 Department of Agricultural Sciences, Università di Bologna, Bologna, Italy

4 International Arctic Research Center, University of Alaska Fairbanks, Fairbanks, Alaska, USA

5 School of Natural Resources and Extension, University of Alaska Fairbanks, Fairbanks, Alaska, USA

6 Institute of Northern Engineering, University of Alaska Fairbanks, Fairbanks, Alaska, USA

References

[1] Chapin FS III, McGuire AD, Randerson J, Pielke R, Baldocchi D, Hobbie SE, Roulet N, Eugster W, Kasischke E, Rastetter EB, Zimov SA, Running SW. Arctic and boreal ecosystems of western North America as components of the climate system. Global Change Biology. 2000;**6**(Suppl. 1):211–223

[2] Serreze MC, Walsh JE, Chapin FS, Osterkamp T, Dyurgerov M, Romanovsky V, Oechel WC, Morison J, Zhang T, Barry RG. Observational evidence of recent change in the northern high-latitude environment. Climatic Change. 2000;**46**:159–207

[3] Hinzman LD, Bettez ND, Bolton WR, Chapin FS, Dyurgerov MB, Fastie CL, Griffith B, Hollister RD, Hope A, Huntington HP, Jensen AM, Jia GJ, Jorgenson T, Kane DL, Klein DR, Kofinas G, Lynch AH, Lloyd AH, McGuire AD, Nelson FE, Oechel WC, Osterkamp

TE, Racine CH, Romanovsky VE, Stone RS, Stow DA, Sturm M, Tweedie CE, Vourlitis GL, Walker MD, Walker DA, Webber PJ, Welker JM, Winker KS, Yoshikawa K. Evidence and implications of recent climate change in northern Alaska and other arctic regions. Climatic Change. 2005;**72**(3):251–298

[4] Intergovernmental Panel on Climate Change. Climate change 2007: The physical science basis. Contribution of working group I to the fourth assessment report of the intergovernmental panel on climate change. iN: Solomon S, Qin D, Manning M, Chen Z, Marquis M, Averyt KB, Tignor M, and Miller HL, editors. New York: Cambridge University Press; 2007. p. 996

[5] Molders N, Romanovsky VE. Long-term evaluation of the hydro-thermodynamic soil-vegetation scheme's frozen ground/permafrost component using observations at Barrow, Alaska. Journal of Geophysical Research. 2006;**11**:D04105

[6] Romanovsky V, Burgess M, Smith S, Yoshikawa K, Brown J. Permafrost temperature records: Indicators of climate change. EOS Transactions. 2002;**80**:589–594

[7] Yoshikawa K, Hinzman LD. Shrinking thermokarst ponds and groundwater dynamics in discontinuous permafrost near Council, Alaska. Permafrost and Periglacial Processes. 2003;**14**:151–160. doi: 10.1002/ppp.451

[8] Smith LC, Sheng Y, MacDonal GM, Hinzman LD. Disappearing Arctic lakes. Science. 2005;**308**:1427

[9] Osterkamp TE. Characteristics of the recent warming of permafrost in Alaska. Journal of Geophysical Research. 2007;**112**:F02S02

[10] Jorgenson MT, Romanovsky V, Harden J, Shur Y, O'Donnell J, Schuur EAG, Kanevskiy M, Marchenko S. Resilience and vulnerability of permafrost to climate change. Canadian Journal of Forest Research. 2010;**40**:1219–1236

[11] Mack MC, Shuur EAG, Bret-Harte MS, Shaver GR, Chapin FS III. Ecosystem carbon storage in Arctic tundra reduced by long-term nutrient fertilization. Nature. 2004;**431**:440-443

[12] Clein JS, Kwiatkowski BL, McGuire AD, Hobbie JE, Rastetter EB, Melillo JM, Kicklighter DW. Modelling carbon responses of tundra ecosystems to historical and projected climate: A comparison of a plot- and global-scale ecosystem model to identify process-based uncertainties. Global Change Biology. 2000;**6**:141–159

[13] McGuire AD, Clein JS, Melillo JM, Kicklighter DW, Meier RA, Vorosmarty CJ, Serreze MC. Modelling carbon responses of tundra ecosystems to historical and projected climate: Sensitivity of pan-Arctic carbon storage to temporal and spatial variation in climate. Global Change Biology. 2000;**6**(Suppl. 1):141–159

[14] McFadden JP, Dugster W, Chapin FS III. A regional study of the controls on water vapor and CO_2 exchange in Arctic tundra. Ecology. 2003;**84**:2762–2776

[15] Hinzman LD, Kane DL. Potential response of an Arctic watershed during a period of global warming. Journal of Geophysical Research. 1992;**97**(D3):2811–2820

[16] McFadden JP, Chapin FS III, Hollinger DY. Subgrid-scale variability in the surface energy balance of arctic tundra. Journal of Geophysical Research. 1998;**103**(D22):28947–28961

[17] Sharratt BS. Observations and modeling of interactions between barley yield and evapotranspiration in the Subarctic. Agricultural Water Management. 1994;**25**:109–119

[18] Sharratt BS. Barley yield and evapotranspiration governed by tillage practices in interior Alaska. Soil Tillage Research. 1998;**46**:225–229

[19] Banimahd SA, Zand-Parsa S. Simulation of evaporation, coupled liquid water, water vapor and heat transport through the soil medium. Agricultural Water Management. 2013;**130**:168–177

[20] Zhuang Q, McGuire AD, Mellilo JM, Clein JS, Dargaville RJ, Kicklighter DW, Myneni RB, Dong J, Romanovsky VE, Harden J, Hobbie JE. Carbon cycling in extratropical terrestrial ecosystems of the northern hemisphere during the 20th century: A modeling analysis of the influences of soil thermal dynamics. Tellus. 2003;**55B**:751–776

[21] Euskirchen ES, McGuire AD, Kicklighter DW, Zhuang Q, Clein JS, Dargaville R, Dye DG, Kimball JS, McDonald KC, Mellilo JM, Romanovsky VE, Smith NV. Importance of recent shifts in soil thermal dynamics on growing season length, productivity, and carbon sequestration in terrestrial high-latitude ecosystems. Global Change Biology. 2006; **12**:731–750

[22] Balshi MS, McGuire AD, Zhuang Q, Mellio J, Kicklighter DW, Kasichke E, Wirth C, Flannigan M, Harden J, Clein JS, Burnside TJ, McAllister J, Kurz WA, Apps M, Shvidenko A. The role of historical fire disturbance in the carbon dynamics of the pan-boreal region: A process-based analysis. Journal of Geophysical Research. 2007;**112**:G02029

[23] Lawrence DM, Slater AG, Romanovsky VE, Nicolsky DJ. Sensitivity of a model projection of near-surface permafrost degradation to soil column depth and representation of soil organic matter. Journal of Geophysical Research. 2008;**113**:F02011

[24] Ruairuen W, Fochesatto GJ, Sparrow EB, Schnabel W, Zhang M, Kim Y. Evapotranspiration cycles in a high latitude agroecosystem: Potential warming role. PLoS ONE. 2015;**10**(9): e0137209. doi:10.1371/journal.pone.0137209

[25] Sitch S, McGuire AD, Kimball JS, Gedney N, Gamon J, Engstrom RN, Wolf A, Zhuang Q, Clein JS, McDonald KC. Assessing the carbon balance of circumpolar Arctic tundra with remote sensing and process based modeling approaches. Ecological Applications. 2007;**17**:213–234

[26] Bittelli M, Ventura F, Campbell GS, Snyder RL, Gallegati F, Pisa PR. Coupling of heat, water vapor, and liquid water fluxes to compute evaporation in bare soils. Journal of Hydrology. 2008;**362**(3):191–205

[27] Starkenburg D, Metzger S, Fochesatto GJ, Alfieri J, Gens R, Prakash A, Cristobal J, Kane D. Assessment of de-spiking methods for turbulent flux computations in high latitude forest canopies using sonic anemometers. Journal of Atmospheric and Oceanic Technology. 2016;**33**:2001–2013. doi: 10.1175/JTECH-D-15-0154.1

[28] Gruber M, Fochesatto GJ. A new sensitivity analysis and solution method for scintillometer measurements of area-averaged turbulent fluxes. Boundary-Layer Meteorology. 2013;**149**(1):65–83

[29] Fochesatto GJ, Mayfield JA, Starkenburg DP, Gruber MA, Conner J. Occurrence of shallow cold flows in the winter atmospheric boundary layer of interior of Alaska. Meteorology and Atmospheric Physics. 2013;127(4):369–382. doi: 10.1007/s00703-013-0274-4

[30] Gruber MA, Fochesatto GJ, Hartogensis OK, Lysy M. Functional derivatives applied to error propagation of uncertainties in topography to large-aperture scintillometer-derived heat fluxes. Atmospheric Measurement Techniques. 2014;**7**:2361–2371. doi:10.5194/amt-7-2361-2014

[31] Bittelli M, Campbell GS, Tomei F. Soil Physics with Python, Transport in the Soil-Plant-Atmosphere System. Oxford, UK: Oxford University Press. 2015, 464 pages. ISBN: 978-0-19-968309-3

[32] Campbell GS, Norman, JM. An Introduction to Environmental Biophysics, second ed., Springer, New York, USA, 1998.

[33] Davin EL, Seneviratne SI, Ciais P, Olioso A, Wang T. Preferential cooling of hot extremes from cropland albedo management. Proceedings of the National Academy of Sciences. 2014;**111**(27):9757–9761

[34] Sharratt BS. Water use, intercepted radiation and soil temperature of skip-row and equidistant-row barley. Agronomy Journal. 1993;**85**:686–691

[35] Mayfield JA, Fochesatto GJ. The layered structure of the winter atmospheric boundary layer in the interior of Alaska. Journal of Applied Meteorology and Climatology. 2013;**52**:953–973

[36] Ortega-Farias S, Carrasco M, Olioso A, Poblete C. Latent heat flux over Cabernet Sauvignon vineyard using the Shuttle worth and Wallace model. Irrigation Science. 2007;**25**:161–170

[37] Ortega-Farias S, Antonioletti R, Olioso A. Net radiation model evaluation at an hourly time step for Mediterranean conditions. Agronomie. 2000;**20**:157–164

[38] Jackson RD. Diurnal changes in soil water content during drying, in field soil water regime. 1973. In: Bruce RR (Ed.). Field Soil Water Regime. 5. SSSA Special Publication. pp. 37–55

[39] Cahill AT, Parlange MB. On water vapor transport in field soils. Water Resources Research. 1998;**43**:731–739

[40] Parlange MB, Cahill AT, Nielsen DR, Hopmans JW, Wendroth O. Review of heat and water movement in field soils. Soil & Tillage Research. 1998;**47**:5–10

[41] Starkenburg D, Fochesatto GJ, Prakash A, Cristóbal J, Gens R, Kane DL. The role of coherent flow structures in the sensible heat fluxes of an Alaskan boreal forest. Journal of Geophysical Research. Atmospheres. 2013;**118**:8140-8155. doi:10.1002/jgrd.50625

[42] Starkenburg DP, Fochesatto GJ, Cristóbal J, Prakash A, Gens R, Alfieri JG, Nagano H, Harazono Y, Iwata H, Kane DL. Temperature regimes and turbulent heat fluxes across a heterogeneous canopy in an Alaskan boreal forest. Journal of Geophysical Research Atmospheres. 2015;**120**:1348–1360. doi: 10.1002/2014JD02233

Chapter 5

Moisture Evaporation from Granular Biopesticides Containing Quiescent Entomopathogenic Nematodes

Carlos Inocencio Cortés-Martínez,

Jaime Ruiz-Vega and

Gabino Alberto Martínez-Gutiérrez

Additional information is available at the end of the chapter

http://dx.doi.org/10.5772/intechopen.68519

Abstract

The moisture evaporation process from granular biopesticides (GBs) containing entomopathogenic nematodes (EPNs) has influence in the shelf-life of these biological products, but the approach to design GBs with desired transport properties lacks of theoretical support to get closer in a better way to formulations design of long-term storage. In this chapter we review the state of art in theoretical studies about the physics of the moisture evaporation to elucidate what are the mechanisms of drying of GBs. We found that several external and internal factors influence the transport process of moisture exchange among others phenomenon that happened in a porous media such as GBs; consequently, complex and highly dynamic interactions between medium properties, transport processes, and boundary conditions result in a wide range of evaporation behaviors. The theory of drying process in two stages for porous materials with high moisture content seems to be a good starting point to explore further the drying of GBs at different scales and mechanistic and correlative models of evaporation are available to analyze the desiccation in different stages of the elaboration process, which is also of interest in the subject area of science and technology of the formulation of EPNs.

Keywords: evaporation, entomopathogenic nematodes, storage stability, reservoir

1. Introduction

Moisture evaporation from porous media is studied by its importance in the drying of foods, building materials and biological products such as biopesticides. The formulation of biopesticides is a process to transform biocontrol agents in a product to exploit their pathogenicity against insect pests. The functions of the formulation oriented towards the biocontrol agents are to improve their storage stability (increased survival time and maintained infectivity), protect them of adverse conditions and potentiate their pathogenicity against insect pests in different developmental stages, which facilitate their transport and use as functions oriented towards the final users [1]. Entomopathogenic nematodes (EPNs) are natural regulators of insect populations and are also applied as biocontrol agents of insect pests in agricultural crops. The infective juvenile (IJ) is the unique free-living stage of EPN; typically, they dwell in the soil, their natural reservoir, until they are able to infect an insect to resume their development and reproduction [2]. The IJs have many strategies to survive when they are subjected to adverse abiotic factors. Particularly, under low moisture conditions in soil, as in drought, they experiment a desiccation process that, when happens as a gradual moisture reduction, allows them to diminish their metabolism gradually until they reach an anhydrobiosis state (life without water) in which they are capable to survive for many years waiting for better moisture conditions to continue their life cycle [2]. Regarding this survival strategy, the granular biopesticides (GBs) as granules or pellets were designed to replicate the desiccation regime of IJs, to gradually arrest their metabolism and to extend their storage stability at room temperature [1].

Each GB can contain up to 3×10^5 IJs and compared with the conventional aqueous suspension it is a better media for long-term storage of IJs at moderate temperature. The main phenomenon that governs the functional performance of the GBs is that if the rate of water reduction from their structure is optimal, the IJs are properly desiccated at a suitable rate and their storage stability is increased [1]. In laboratory, it has been found that the removal of moisture content from pellets by evaporation is well related to the survival of the IJs (Pearson $r = 0.725$) and infectivity on *Galleria mellonella* larvae (Pearson $r = 0.904$), suggesting that the relationship between the pellet drying rate and the storage stability may be an important factor to improve the shelf life of GBs containing *Steinernema glaseri* IJs [3].

On the one hand, the rate of water reduction of IJs to survive to desiccation is influenced by their size, energy reserves, metabolism, genetic, historical adaptations, origin and other characteristics that are unique to each EPN species and even strains. Also, as a living entity, they receive stimulus from the ambient biotic and abiotic factors. Knowledge of EPNs biological fitness is needed for their optimal formulation, particularly their desiccation tolerance (survival under water evaporation at low relative humidity or hypertonic osmotic conditions) and also to evaluate the possibility of improvement through pre-adaptation, selective breeding, genetic engineering or others methods [4]. On the other hand, the drying process of the formulated must be optimal to induce the IJs into an anhydrobiotic state. To produce an EPN product with best performance, a proper selection of carrier's materials, adjuvants, elaboration process to produce them and a certain combination of ambient storage conditions that allow the evaporation of enough moisture from the GB at a rate reduction particularly suitable to gradually remove the water from the IJ's body is required [1, 5]. However, it is worth mentioning that the materials selection and its combination to achieve the desired drying characteristics of the GBs are carried out by the formulator in an iterative way until the proper solution is found.

Actually, both research lines (EPN biological fitness and optimal drying of GBs) are ongoing, but the second is less common, although the major factor to enhance EPN longevity and, perhaps, increase the range of applications, given the inherent limitations of EPNs survival ability, it is likely to be the improvement of the formulation [4] and theoretical literature concerning energy and mass transfer is extensive. In this regard, if the mechanisms of transport of moisture, temperature and oxygen in these GB were understood and focused on the transport properties of the materials formulation based on the materials science approach, the extensive empirical knowledge in formulation of EPNs would be complemented with theoretical support. It is suitable for the design of these biological products because it can help to build mathematical models that describe the migration patterns of moisture, the thermodynamic equilibrium and exchange of oxygen [6], and would serve for analytical solving of design problems. Afterwards, several methods could be tested to optimize the moisture removal in the formulated product. Also, modelling could help to avoid too long, unnecessary or expensive experiments on the EPN-GB system.

Recently, it has been found that the drying kinetics of pellets is reproduced by a surface evaporation model [7] with a percent relative average deviation value of 21.95%; consequently, there is room for improvement through the proposed simplifying assumptions, but they need to be determined experimentally and expressed in mathematical form to feedback the model [3]. But, to improve this theoretical approach, it is necessary as first step to understand the water migration by evaporation from GBs towards the surrounding atmosphere and its variation factors. Hence, the objective of the present work is to make a review and discuss the state of the art in theoretical studies about the physics of moisture evaporation process from porous media to elucidate what are the basic mechanisms of moisture migration from GB, which is of paramount interest in the subject area of science and technology for the formulation of EPNs.

2. Problem statement of moisture evaporation from granular biopesticides

The pellets are solid particles in the form of spherical granules of 10–20 mm in diameter (**Figure 1**). Drops of a suspension of nematodes are dropped on a powdery material to entrap the IJs into a solid matrix, which results in the formation of the granules or pellets. The powder material may be one, or a combination of, several different carrier materials and adjuvants to produce GBs with the desired characteristics [1, 5]. Experimental wettable powder formulations have been reported to include inert powders such as talc, sand, diatomaceous earth and various clays. Natural ingredients are often used in formulations in order to maintain the

Figure 1. Various diatomaceous earth pellets containing *S. glaseri* IJs.

environmentally "green" concept associated with biopesticides. Polymers are a usual material for pelletisation of EPNs and often include natural carbohydrate and/or protein polymers such as starch, sodium alginate, acacia gum, lignin and gelling agents amongst others [6, 8, 9].

The nematodes can be divided into two groups, slow-dehydration strategists and fast-dehydration strategists, depending upon the rates of water loss in the environment that they will survive. *S. carpocapsae* IJs survive well at high rate of water removal and *S. glaseri* IJs survive better at low rate of water removal [10, 11]. Therefore, the major cause of nematode survival to desiccation in anhydrobiosis is the controlled rate of water removal from the IJs [12]. Therefore, the quiescent state of IJs and their shelf life in GB are strongly influenced by moisture content [12–14].

In laboratory, the GBs are stored in closed room where evaporation occurs in calm air, and they are subject to artificial variations to replicate the shelf storage conditions and to test their storage stability over time (temperature from 20 to 30°C, relative humidity of 0–100% and no wind-flow present). It has been observed that the initial moisture content of GBs containing EPNs is usually between 40 and 100%, from which a part can be removed faster in few days or slowly in various months. In the most successful case, *S. carpocapsae* IJs survived up to 7 months (**Figure 2**) and maintained their infectivity on *G. mellonella* above 70% in water dispersible granules (WDG) stored at 25°C, whose moisture is reduced at a rate of 8.26 $\times 10^{-7}\% \, s^{-1}$ [13, 15]. In the less successful case *S. glaseri* IJs survived 8 days and maintained its infectivity on *G. mellonella* above 80% in diatomaceous earth pellets (**Figure 3**) stored at 23 $\pm$ 3°C, whose rate of pellet moisture reduction is $6.134 \times 10^{-5}\% \, s^{-1}$ [3]. This reduced shelf life at room temperature limits their commercial exploitation, especially if GBs contain *S. glaseri* IJs and if these are compared with the long shelf life of commercial chemical pesticides.

http://dx.doi.org/10.5772/intechopen.68519

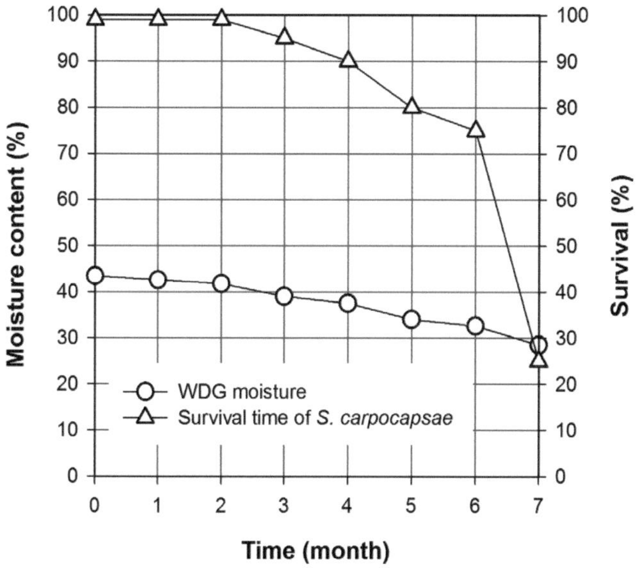

Figure 2. Mean survival of *S. carpocapsae* IJs and moisture content of the WDG formulation stored at 25°C, reported in [13].

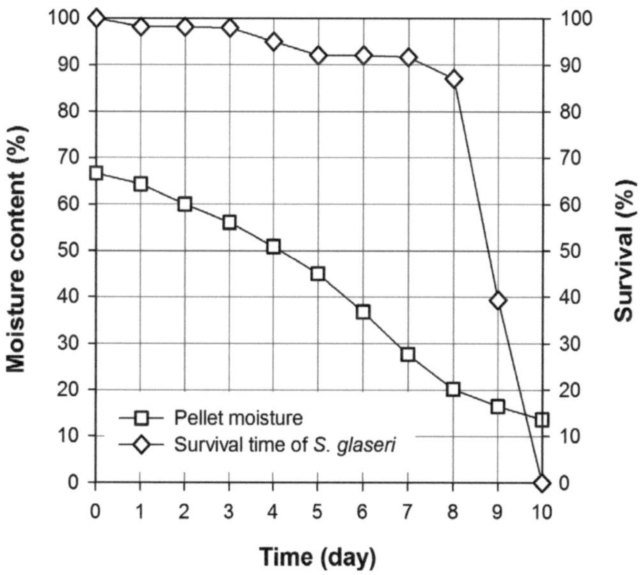

Figure 3. Mean survival of *S. glaseri* IJs and moisture content of the diatomaceous earth pellet formulation stored at 23 ± 3°C and high RH (96%), reported in [3].

It is thought that the desiccation process of IJs happened by slow absorption of the aqueous suspension containing them for the carrier material, starting a reduction process of moisture, which developed at an appropriate rate, and diminished the metabolism of the IJs [5]. In fact, 52% of the variation in its survival rate is explained by the behaviour of the moisture content of the DE pellet, whereas 84% of the variation in infectivity on *G. mellonella* is explained by the survival of *S. glaseri* IJs in diatomaceous earth pellets [3]. The hypothesis is that the sudden death of the IJs formulated under these conditions is due to the diffusional migration of water molecules surrounding the IJs to the DE pellet surface followed by the contact of the DE particles with the nematode cuticle which absorbs moisture faster from the IJ's cells [3]. Also as it can be observed in **Figure 4**, the behaviours of drying kinetics of diatomaceous earth pellets are different with or without *S. glaseri* IJs.

Recent inspection of cross-sectional area of diatomaceous earth pellets using scanning electron microscopy showed particles as plate-form and non-uniform pore distribution that form a complex and disordered microstructure [6], probably dominated by a double porosity due to two distinct distributions, one for the region of macroscopic porosity between particles, and another for the region of microscopic porosity within particles [16]. Due to the above-mentioned facts, in next sections, we will be dealing with theories applied to understand the moisture evaporation from porous media to understand how the evaporation from pellets happened, which can be useful to set design criteria to elaborate GB reservoirs for the optimum storage of EPNs.

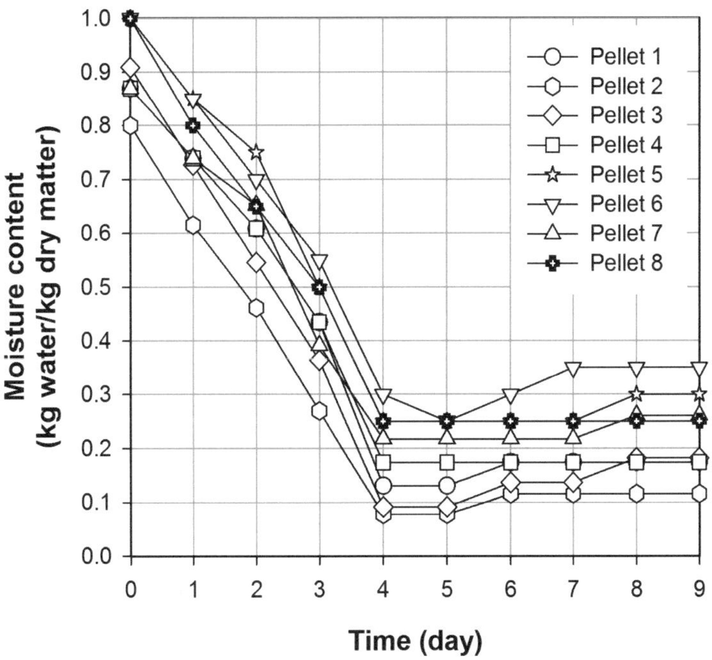

Figure 4. Drying kinetics of diatomaceous earth pellets without EPNs with several initial moisture contents, stored in quiescent surrounding at room temperature (23 ± 3°C).

3. Theoretical background

3.1. The evaporation process in porous media

In the evaporation process, liquid water is transformed to water vapour and transferred from the surface of evaporation to the surrounding atmosphere, while a porous medium is a material with a skeletal solid structure with interconnected void spaces that allow fluid to pass through the medium and is mainly characterized by its porosity, i.e. the ratio of the void space to the total volume of the medium [17–19]. The permeability is a measure of the flow conductivity in the porous body and the tortuosity represents the hindrance to flow diffusion imposed by local boundaries or local viscosity; both these are important characteristics for the combination of the fluid and structure of the porous medium, respectively [17].

In porous media, evaporation involves mass and energy transport including phase change, vapour diffusion and liquid flow, resulting in complex displacement patterns affecting drying rates [20] and is the only one mechanism by which moisture can leave the GBs. Moreover, at the pore scale several mechanisms influence the macroscopic behaviour of the drying process. Among others, the phase change at the liquid-gas interface, diffusion and convection of mass and heat, type of flow and the effect of the combined viscous, capillary and buoyancy forces on the receding of the liquid-gas interfaces are also the influencing factors in drying process [21]. Most evaporation processes are assumed to be viewed as transport at the pore space where water is displaced by air; this moving zone between wet and partially dry zones is named the drying front in [22].

According to the theory of drying in two stages, drying rates from an initially saturated porous medium often exhibit distinct transitions whereby initially high drying rate (denoted as stage I or constant rate period) abruptly drops to a lower rate (stage II or falling rate period) governed by vapour diffusion through the porous medium [20, 23]. The drying stage 1 occurs by convection on the material surface at an evaporation rate relatively high and nearly constant. The factors involved in its development are the external conditions of mass and heat exchange in the system related to the properties of the surrounding air (temperature, pressure, humidity, convective airflow velocity and area of the exposed surface) and supplied by capillary liquid flow connecting a receding drying front with an evaporating surface [24].

The drying stage 2 starts when the moisture content of the porous media is less than its critical moisture content. At a certain drying front depth, continuous liquid pathways are interrupted when gravity and viscous effects overcome capillary forces in porous media and simultaneously the moisture content is less than its critical moisture concentration. These events marks the onset of a new evaporation regime with lower evaporation rate limited by transport properties of porous media, known as the drying stage 2 [22]. In this period the process is influenced by internal factors related to the properties of the material such as water activity, internal structure (porosity and tortuosity), chemical composition and transport properties (thermal and hydraulic conductivities, moisture diffusivity and vapour diffusion) [22–24]. The rate of drying in the two different periods has mainly been modelled numerically from a set of equations taking into account vapour and liquid transfers along with boundary conditions.

In [25], the convective drying is considered as a process of three successive steps. The first step is liquid movement in porous media from the wet interior to the gas-solid interface across internal pore, particle surface, etc. This step is slower for larger solids and/or low moisture concentration in the porous material. The second is evaporation due to heat by energy supplied to change liquid into vapour. The third step is vapour movement to the surrounding gas by diffusion and convection. In general, different stages are often used for diagnostics and classification of evaporative drying of porous media, which depend on the focus of the studies and the interpretation of the rate limiting processes. Consequently, complex and highly dynamic interactions between medium properties, transport processes and boundary conditions result in a wide range of evaporation behaviours [20] and internal and external factors are common influences on most of the evaporation behaviours.

3.2. External mass transfer

3.2.1. *Classic evaporation theory*

A brief history of the theories of evaporation is presented in reference [26]. The Dalton's assay was one of the major events in the development of the evaporation theory and states that the rate of water evaporation is proportional on the saturation deficit of the air, which is given by the difference between the saturation vapour pressure at the water temperature, e_w and the actual vapour pressure of the air e_a [27, 28]:

$$E_L = C(e_w - e_a) \tag{1}$$

where E_L = rate of evaporation (mm/day) and C = constant; e_w and e_a are in mm of mercury. Eq. (1) is known as Dalton′s law of evaporation. Also, this proportionality is affected by the wind velocity, the RH and the moisture concentration in the solid surface; therefore, the moisture will evaporate from the surface until the surrounding air saturates [28]. The similarity theory is a standard basis for predicting evaporation rate from a free water surface and states that convective heat and mass transfer are completely analogous phenomena if the mass flow from the surface is caused by diffusion, which requires that the content of the diffusing species be low and the diffusional mass flux is low enough that it does not affect the imposed velocity field. However, these two analogies are not valid if applied to a capillary porous media containing a liquid [28].

3.2.2. *The constant rate period*

According to the theory of two stages of drying of porous media, during the 1-stage, the rate of water loss per unit surface area remains nearly constant and close to evaporation rate from free water surface and is attributed to the persistence of continuous hydraulic pathways between the receding drying front and surface of porous media where liquid flow is sustained by hydraulic gradient towards the evaporation surface [29]. Numerous experimental and theoretical studies have established the existence of such a constant phase during drying of porous media, typically under mild atmospheric demand [20, 30].

The intrinsic characteristic length L_C, a measure of the extent of hydraulic continuity and the strength of capillary driving force deduced from pore size distribution of a medium that control the transition from liquid-flow–supported stage 1 to diffusion-controlled stage 2 during evaporation from porous media is proposed in [20]. The L_C is used for predicting the end of stage 1 evaporation and considering balance between gravitational, capillary and viscous forces. The theory assumes that in the stage-1, air first enters the largest pores in a complex porous medium while menisci in smaller pores at the evaporating surface may remain in place (albeit with decreasing radii) of curvature. Thereby, liquid moves into the pore space (large size pores with lower air-entry value to smaller size water-filled pores) and then towards the surface of the body where the moisture evaporation happens; according to [20], such mass flow may be sustained as long as capillary driving forces are higher than gravitation and viscous forces. In [20], characteristic lengths for evaporation from porous media are theoretically developed for the following three conditions:

1. pore size distribution and gravity characteristic length L_G,

$$L_G = \frac{1}{\alpha(n-1)}\left(\frac{2n-1}{n}\right)^{(2n-1)/n}\left(\frac{n-1}{n}\right)^{(1-n)/n} \tag{2}$$

2. pore size distribution effects on the viscous characteristic length L_V,

$$L_V = \frac{K(\theta)}{e_0}\Delta h_{\mathrm{cap}} \tag{3}$$

3. combined gravity and viscous length L_C,

$$L_C = \frac{L_G}{1-\frac{e_0}{K(\theta)}} \tag{4}$$

where $K(\theta)$ is the unsaturated hydraulic conductivity, α is the inverse of a characteristic pressure head, n is the pore size distribution, e_0 is the water flow supporting evaporation rate and Δh_{cap} is the maximum capillary driving force. This theoretical approach is based on experimental data of evaporative drying of two quartz sand media with particle sizes ranging from 0.1 to 0.5 mm (denoted as "fine sand") and from 0.3 to 0.9 mm ("coarse sand") to quantify the evaporation rates from sand-filled cylindrical Plexiglas columns of 54 mm in diameter and 50–350 mm in length and rectangular Hele-Shaw cells 260 mm in length, 10 mm in thickness and 75 mm in width with a top boundary open to the atmosphere. As expected, the highest initial evaporation rate corresponds to high temperature and low humidity (28°C and 31% RH), whereas the lowest evaporation occurred for cool and humid conditions (21°C and 58% RH).

3.2.3. Natural convection theory

Natural convection is defined as air movements brought about by density differences in hot and cool air, whereas forced convection is the movement of air brought about by an external force. In natural convection, fluid motion is due to gravity that creates a buoyant force within the fluid, which lifts the heated fluid upward. Since the fluid velocity associated with natural convection is relatively lower than those associated with forced convection, the corresponding convection transfer rates are also smaller [31]. The Rayleigh number that measures the intensity of natural convection, based on the macroscopic length scale L, is defined in reference [19] as

$$Ra_L = \frac{g\beta\Delta TL^3}{v_f\alpha_f} \tag{5}$$

where g is the gravity constant, a_f is the thermal diffusivity of fluid, β is the volumetric temperature expansion coefficient, L is the macroscopic scale length scale, ΔT is the temperature scale and v_f is the kinematic viscosity of fluid. Based on natural convection theory, heat and mass transfer from porous media in quiescent fluid environments have been extensively studied by its importance in the design or performance of the systems when it is desirable to minimize heat transfer rates or to minimize operating costs [32].

3.3. Internal mass flow in porous media

The internal process of mass transfer during drying is usually described using a convective model (known as a capillary-porous model) based on fluid pressure gradients (Darcy's law) and techniques of scale change as a representative elementary volume (REV) in order to express the transition from a microscopic level to a macroscopic one in the conservation equations and a diffusive model based on the gradients of moisture concentration, a phenomenon described by Fick's law. This model is a simplification of the capillary porous model if the drying is supposed isothermal with no gravity effects in the solid-liquid water system [33–35].

3.3.1. Fickean diffusion

The diffusion is known as the preponderant internal mass transfer mechanism during drying of porous media. According to Fick's first law, the matter flows erratically from moist regions towards dry regions inside bodies, assuming that the moisture gradient is the unique driving force of the flow [37]. However, although several empirical equations have been proposed to predict mass transfer on this basis, much more must be explained [37]. For instance, this laws make several assumptions and simplifications that are often unrealistic to model water diffusion during drying as materials are non-heterogeneous and isotropic media and diffusion coefficients are not correlated to moisture content; samples are in most cases considered as having regular shapes; heat transfer during drying is disregarded and collapse of vegetable tissues by water loss is also neglected [36, 37].

The effect of microscopic structure on mass transfer has been completely discarded, but a disordered internal geometry caused by the percolation phenomena is very common in the structure of most porous media, sometimes described by the fractal geometry. The complex microstructure affects the water diffusion phenomena, resulting in anomalous diffusion and involving complex parameters such as fractal dimension and spectral dimension [37]. The inclusion of a porous medium affects the forms of Fick's first and second laws for diffusion of chemical species in aqueous solution in two general ways: first, the existence of the solid particles comprising the porous medium results in diffusion pathways that are more tortuous. This increased tortuosity reduces the macroscopic concentration gradient and, therefore, reduces the diffusive mass flux relative to that which would exist in the absence of the porous medium as in single droplets. Second, there may be interactions between the diffusing species and the solid porous media that either directly affect the mass of the diffusing species in aqueous solution (e.g. sorption) and/or result in physicochemical interactions that affect the tortuosity [18]. All forms of Fick's first and second laws for governing macroscopic diffusion through porous media include an effective porosity, ε_{ff} and a mass diffusion coefficient D [18].

3.3.2. Capillarity

The migration of the liquid phase in a deformable porous matrix, by convective transport, is managed by the generalized Darcy's law [33–35]. Darcy's law in its simplest form expresses the proportionality between the average velocity $\underline{\mathbf{v}}$ of a fluid flow and the flow potential, comprised by the pressure gradient Δp existent through porous media and the gravitational contribution and is applicable to multiphase mixtures as opposed to Fick's law, which requires the assumption of a homogeneous mixture [19]. The proposed relationship is as follows:

$$\mathbf{v} = k\frac{\Delta p}{\Delta x} \tag{6}$$

in this expression, k is the hydraulic conductivity and describes the ease with which a fluid can flow through the pore spaces. Darcy's equation is valid for incompressible and isothermal creeping flows. In a complex form of Darcy's law, the rate of flow is related to the pressure gradient in the liquid, $\nabla\langle V_l\rangle^l$ [33]. Eq. (7) expresses the relationship between the liquid-phase velocity $\langle V_l\rangle^l$ and solid one $\langle V_s\rangle^s$ as follows:

$$\langle V_l\rangle^l = \langle V_s\rangle^s - \frac{\overline{\overline{K}}}{\varepsilon_l \mu_l}\cdot(\nabla\langle P_l\rangle^l) \tag{7}$$

where $\overline{\overline{K}}$ is the permeability of the porous medium, ε_l is the liquid fraction, μ_l is the liquid dynamic viscosity and $\Delta\langle P_l\rangle^l$ is the gradient of fluid pressure.

3.3.3. The falling rate period

The stage 2 is governed by vapour diffusion through the porous medium. This period is divided in a first decreasing-rate period (FDR) characterized by breaks in the uniformity of water content close to the surface and slight augmentation of the heat surface; and a second decreasing-rate period (SDR), correlated with a discontinuous liquid network into the porous medium and with the development of a dry receding front from the free surface of the porous sample [38].

The falling rate period is expected to be short for two reasons: (1) the liquid mass corresponding to the films is small compared to the mass of liquid initially present in the medium and (2) the external mass transfer length scale (the mass external boundary layer typically) is typically greater than the thickness of the medium, which implies that the mass transfer resistance due to the receding of film tips within the medium is weak compared to the external mass transfer resistance [39].

3.4. Mathematical models of moisture evaporation from porous media

The drying behaviour of porous material can be described with a model and the porous media is described in multiple length scales [40, 41]. The macroscopic length scale is defined by the overall physical domain indicated by the length scale L. The microscopic length scale captures the detailed morphology and is indicated by d. A REV is defined as a volume whose size L_{REV} lies between length scale d and L, i.e. $d \ll L_{REV} \ll L$ [19]. Thus, the modelling of drying process of porous media can be developed at different scales in a process that should be initiated typically to the pore-scale and then model it at larger and larger scales up to dryer scale or product scale in the case of designs of dryers.

The macroscopic variables of the drying process are commonly defined by the volume average of the microscopic variables over the REV and the nature of the product is the result of a diversity of multiple factors and their relationships among themselves that increases the complexity of the evaporation process [19, 41]. In the formulation of EPNs, the interesting scale is the product scale, being one single granular biopesticide. The continuum approach and pore network models are approaches for the modelling of the drying process of the porous media [41].

3.4.1. The continuum approach

In the continuum approach, variables (e.g. temperature) are averaged over the volume, the REV. Equations for the conservation of liquid, air and energy are supplemented with boundary and initial conditions. The continuum approach can be solved by efficient numerical techniques at a large scale in comparison to the pore scale, which is a great advantage. The effective parameters, such as vapour diffusivity, permeability, thermal conductivity and capillary pressure have to be determined by dedicated experiments. The continuum approach fails when the pores are large compared with the system and is not able to easily take structural features of the medium into account. Moreover, the computation of the effective properties at the scale of a REV is necessary in the continuum approach [41]. The parameters of the continuum model can be assessed for a certain pore structure using a pore network model.

3.4.2. The pore network approach

Porosity is a primary property of the granules, which dominates a wide range of secondary properties as water flow or air entrapment, both strongly linked to the microstructure [42, 43]. Pore network models are based on a porous structure represented as a network of pores and throats and these models can be used to simulate the drying process at the pore level because they can take into account important features of the microstructure as the role of large pores and their distribution on motion of the gas-liquid menisci in the pores, diffusion, viscous flow, capillarity and liquid flow [40, 41, 44]. The developing of models that permit to analyse the influence of the porous microstructure has at least two motivations. One is the computation of the effective parameters at the scale of a representative elementary volume REV of the microstructure. A second one is to analyse drying at the scale of the product without assuming a priori existence of the REV that is associated with the continuum approach. Pore network models have been used in both cases and have been described in two and three dimensions [44].

3.4.3. Pore network models

A pore network model for the evaporative drying of macroporous media was presented in [30]. The model takes into account the heterogeneity of the pore size distribution and the pore wall microstructure, expressed through the degree of pore wall roundness for viscous flow through liquid films, gravity, and for mass transfer, both within the dry medium and also through a mass boundary layer over the external surface of the medium. The model is used to study capillary, gravity and external mass transfer effects through the variation of the three dimensionless numbers: a film-based capillary number Ca'_f, that expresses the ratio of viscous forces to capillary forces in the films;

$$Ca'_f = \frac{4\mu_1 D_M C_e}{\rho_1 \gamma \bar{r}_t} \tag{8}$$

the Bond number, Bo, that expresses the ratio of viscous forces to capillary forces in the films;

$$Bo = \frac{g_x \rho_1 \bar{r}_t^2}{\gamma}, \tag{9}$$

and the Sherwood number, Sh, that describes mass transfer conditions within the mass boundary layer over the product surface.

$$Sh = \frac{\lambda}{\delta} = \frac{\lambda \bar{r}_t}{\Delta}, \tag{10}$$

where $\lambda = \frac{D_{eff,s+}}{D_{eff,s-}} > 1$ is the ratio of external to internal effective (volume-averaged) diffusivities, $\delta = \Delta/\bar{r}_t$ is the dimensionless value of the mass boundary layer and Δ is its corresponding thickness. g_x is the gravity acceleration component in the flow direction, μ_1 and ρ_1 are the liquid-phase viscosity and density, γ is the interfacial tension, $\bar{r}_t$ is the average throat radius within the porous medium, D_M is the molecular diffusivity and C_e is the equilibrium (at vapor pressure) concentration of the volatile species. The film and dry pore regions are coupled through mass conservation at the front evaporation in a single scalar variable, Φ, mass transport through both the film and dry regions:

$$\Phi = \frac{J(\rho) - \mathrm{Bo}Ix(\xi) + \mathrm{Ca'}_f\zeta}{J_p + \mathrm{Ca'}_f} \tag{11}$$

where the variable Φ is subject to the following boundary conditions; at the percolation front P, where the films emanate as $\zeta = 1$ and $\rho = 1$, at the evaporation front I as $\zeta = 1$ and $\rho = p$ and at the top of the mass boundary layer as $\zeta = 0$ and $\rho = 0$. The effect of gravity is analysed for two cases, when it is opposing and when it is improving drying. For the second case, a two-constant rate period evaporation curve was found when viscous forces are strong and mass transfer in the dry region is fast enough compared to gravity forces. Especially in this regime, water flow is driven primarily by gravity to compensate for evaporation occurring at the film tips [30].

The incorporation of gravity can be done by considering a well-chosen invasion throat potential dependent on variables such as its width of the throat, the relative position in the gravity field and the Bond number [30]. If thermal effects are not included, the pore network model requires to be coupled with mass transfer at the open surfaces and under isothermal conditions. But certainly, temperature has an effect on viscosity and surface tension of the liquid and vapour diffusion coefficient, among others. Moreover, the temperature gradient affects the drying process and distribution of moisture during its development. The effect of heat and mass flow on the drying process in simulation has been reviewed recently in [45].

3.4.4. *The diffusion model*

Until now, the unique effort to study the moisture migration from GB was the application of a classical temporal surface evaporation model (Eq. (12)) of Crank [7] based in the Fick's second law to calculate the moisture content of a diatomaceous earth pellet at any given time in the drying process [3]. In this model, differences at initial and final concentration are the driving force of change of moisture in time of a sphere and to take into account all relevant physical transport mechanism in the drying process all effects are lumped on the diffusion coefficient, implying that the detailed effects of the pore microstructure are ignored.

$$\frac{MC(t) - MC_1}{MC_1 - MC_0} = \sum\nolimits_{n=1}^{\infty} \left[\frac{6 \cdot L^2 \cdot exp^{(-\beta_n^2 Dt/R^2)}}{\beta_n^2\{\beta_n^2 + L \cdot (L-1)\}} \right] \tag{12}$$

The use of a large number of series term in Eq. (12) makes their practical use difficult. Also, this approach model is limited to the two phases' system (solid and liquid) and the experimental determination of the effective diffusion coefficient is difficult because of its variation in space [3]. The evaluation of moisture diffusivity using numerical techniques has become a usual methodology in recent years [46] and these numerical methods seem to be a powerful tool for the researchers in formulation of EPNs in GBs.

3.4.5. *Drying-strain relation*

Studies for optimal control of drying of porous media focused on the assessment of drying effectiveness were found in the literature review. Although for now the use of dryer technology where GBs could be properly dried is non-existent, these approaches are of interest for the formulation process of EPNs. The work of Kowalski and co-workers [47] is dedicated to numerical simulations of optimal control applied to saturated capillary-porous materials subjected to convective drying based in a thermo-hydro-mechanical model. The differential equation expressed in terms of strains, temperature and moisture content is as follows:

$$[2M\varepsilon_{ij} + (A\varepsilon + \gamma_T\vartheta - \gamma_X\theta)\delta_{ij}]_{,j} + \rho g_i = 0 \tag{13}$$

where K and M are the elastic shear and bulk modulus, ε is the volumetric strain, ρ is the mass density of the body, g is the gravity acceleration (neglected in further considerations), T is the temperature, X is the moisture content and the index (j) denotes differentiation with respect to coordinate $j = \{x, y, z\}$. $A = (K - 2M)/3$, $\gamma_T = 3Kk^{(T)}$, $\gamma_X = 3Kk^{(X)}$. Here, $k^{(T)}$ and $k^{(X)}$ are the respective coefficients of linear thermal and humid expansion. Eq. (13), after differentiating with respect to the coordinate j and using the tensor of small strains expressed by the derivative of displacement u_i, becomes

$$\varepsilon_{ij} = \frac{1}{2}(u_{i,j} - u_{j,i}) \tag{14}$$

which is the displacement differential equation, jointly with the differential equations expressing liquid concentration and temperature supplemented with appropriated initial and boundary conditions that allow to realize numerical estimations of the drying kinetics and deformations of the drying media, and by implication, of the drying-induced stresses. The whole set of differential equations were initially elaborated for the 2-D geometry of a cylindrical shape [47]. The optimization procedure is illustrated on the kaolin-clay material in the form of cylindrical samples (40 mm in radius and 40 mm in height) and the genetic algorithm method was used to simulate the optimal work of the dryer. The authors conclude that drying rates are accelerated if the drying induced stresses are small, and slowed down if the stresses tend to overcome the strength of the material. The formulation of mathematical optimization procedure based in such rigorous principles of rational control of drying and their experimental validation are useful to find optimal drying processes of porous media [47].

3.4.6. *Drying of droplets*

Drops are subject to theoretical and experimental analysis to determine how the moisture is lost under different conditions [41, 48–56]. Theoretical models for the drying of single droplets are of interest for applications of formulation of EPNs since GBs can be made by liquid penetration in powders or can contain soluble and insoluble adjuvants subject to evaporation. The theoretical study of evaporation of liquid droplets on solid substrate is based on the assumptions of diffusion-controlled mass transfer in the gas phase, constant temperature over the whole system (isothermal conditions) and neglects the effect of convection in the vapour phase [41]. However, when the thermal effects due to evaporative cooling in the classic model are introduced, the results show that the evaporation slows down by increase of the latent heat of evaporation and the substrate thickness as well as by a decrease of the substrate thermo-conductivity. The theoretical predictions using this model do not have good agreement if the substrate temperature deviates from the room temperature and the possible reason of this deviation is the increasing importance on thermal-buoyancy convection at higher temperatures [54].

The regular regime method is useful to determine the content-dependent diffusion coefficient for systems in which the relation among moisture diffusivity and moisture content are lineal and the last one decreasing below the critical moisture concentration or also for situations where the drying rate is dominated by mass transfer inside the drying specimen [57]. Under this theory, the drying curves show an induction period in which the drying rate is conditioned by the moisture distribution at the beginning of the drying process and a regular regime period in which the drying rate is not correlated and thereby of the moisture distribution at the beginning of the process. The establishment of the moisture range at which the regular regime occurs during the isothermal drying is the condition to apply this method, particularly for a given material and conditions, the drying curves will converge in a regular regime curve, even for different moisture contents in a single curve named the regular regime curve [49].

The objective of the work of [49] was to develop a method for quantitatively calculating the effective moisture diffusivity of isothermally dried biopolymer drops and to acquire activation energy to be used as a discriminating parameter for selecting effective wall materials against lipid oxidation. The biopolymer´s effective moisture diffusivity was dependent on moisture content and temperature. Therefore, air temperatures must be lower than 80°C for an appropriate analysis of the water diffusion mechanism using the regular regime methodology. Also, the activation energy provides a quantitative measurement for selecting potential good wall materials against lipid oxidation [49].

In other approach, the interactions of droplets during its deposition on porous material for the agglomeration of particles by spray-fluidized process were studied by [55]. In this work, the penetration of liquid into the porous layer was assumed to be governed by Darcy's law. In reference [48] the molecular kinetic theory was used to model the droplet spreading, and Darcy's law to describe the one-dimensional liquid penetration into the substrate. Particularly, this approach is of interest for applications of formulation of EPNs because the elemental principle of the methods of formation of GBs is to deposit droplets of aqueous suspension containing EPNs over a layer of a mixture of material (carrier and adjuvants) and then the penetration of the liquid carrier happened [5, 9]. After that, the materials are mixed and compacted in different ways (i.e. agitation, eccentric rotational motion and compaction by rolling, among others). These are reasons to optimize the capillary penetration of aqueous suspension into the porous layer; the initial moisture content and moisture evaporation are critical factors in the design of granules for storage and transport of EPNs for biological applications [3, 50]. Following this approach, in [56] was developed a method to quantitatively describe the evaporation effect on radial capillary penetration of liquids in thin porous layers.

3.4.7. Natural convection in cavities

The case study of Prakash et al. [32] is moisture migration in a rectangular cavity (2 m in height and 1 m in diameter) with half the cavity filled with silica gel and their paper presents a general method of solving heat and moisture transfer by analysing a two layer system with a fluid overlying a hygroscopic porous medium where turbulence in the fluid layer affect the natural convection flow. Also, these approaches are of interest for applications of formulation of EPNs if the phenomenon is scaled adequately to small containers in which GBs are deposited for storage, transportation and commercialization, because they permit air in the container to contact the exposed surfaces of the packages for oxygen exchange with EPNs [58, 59]. The model is based in equations for fluid flow and heat transfer and equations for moisture migration in porous media. The model equations were discretized using the control volume formulation and solved using the SIMPLE algorithm. The model is capable of simulating flow only when turbulence in the porous medium can be considered to be negligible. However, this would not be the case for porous media of high permeability. In order to overcome this limitation, the authors suggest that a turbulence model for the porous medium needs to be incorporated into the existing model.

4. Concluding remarks

According to this review focus, it has been found that the evaporation theory from porous media is well developed, and is affected by temperature, atmospheric pressure, humidity, water quality, topology, size and organization of particles and pores, fissures, type of elaboration process and shape of surface among others in an interdependent form. Therefore, as a design object, GBs are the porous granular structure whose key evaporative processes are subject to multiple extra- and intra-structure factors that affect the way of water removal and even other physical, chemical and mechanical processes in a complex way. For the above, considering all the phenomena in the design process of GBs without theoretical support and mathematical tools for the analysis of the transport processes limits the design of optimized solutions. Nevertheless, it is reasonable to think that some improvements can be achieved on this topic because the optimization methods (as the topology optimization procedure used to design heterogeneous materials and stochastic optimization methods employed to reconstruct or construct such microstructures gave limited but targeted structural information) provide a systematic means of designing of microstructures with tailored properties for a specific application. Also, mechanistic and correlative models of evaporation are available to analyse the desiccation of GBs, i.e. describe most of the moisture evaporation process from GBs that could be useful to estimate the time required to reach optimum moisture or conditions to achieve the adequate drying rate of different EPNs. The theory of drying processes in two stages for porous materials with high moisture content seems to be a starting point to explore further the moisture evaporation from GBs at different scales. We suggest that considering the complex physical, chemical and mechanical processes carried out in soil to supply the necessary conditions (i.e. oxygen, humidity, temperature, mechanical strength) for the nematodes to achieve their survival and persistence in soil, and the actual development of the research area, new characteristics should be accomplished in the formulates to mimic the anhydrobiosis induction processes and maintain the stationary state of quiescence for the successful long-term storage of viable EPNs, as if the formulates were a microstructured porous reservoir. This also implies previous work to improve the EPNs biology fitness, e.g. on the pre-acclimatization of IJs before and after formulation. For which, the use of desiccators' technology to gradual drying of large amounts of IJs is a potential topic not commonly explored in the formulation process of EPNs.

Acknowledgements

This work was supported by the SIP project number 20170139. We thank to the Instituto Politécnico Nacional (IPN) and to the Comisión de Operación y Fomento de Actividades Académicas (COFAA) from IPN.

Author details

Carlos Inocencio Cortés-Martínez*, Jaime Ruiz-Vega and Gabino Alberto Martínez-Gutiérrez

*Address all correspondence to: solemia7@hotmail.com

Instituto Politécnico Nacional, CIIDIR Oaxaca, Oaxaca, México

References

[1] Grewal PS, Peters A. Formulation and quality. In: Grewal PS, Ehlers RU, Shapiro-Ilan DI, editors. Nematodes as Biocontrol Agents. 1st ed. Wallingford: CABI; 2005. pp. 79-90

[2] Glazer I. Survival biology. In: Gaugler R, editor. Entomopathogenic Nematology. 1st ed. Wallingford: CABI; 2002. pp. 169-187

[3] Cortés-Martínez CI, Ruiz-Vega J, Matadamas-Ortiz PT, Lewis EE, Aquino-Bolaños T, Navarro-Antonio J. Effect of moisture evaporation from diatomaceous earth pellets on storage stability of *Steinernema glaseri*. Biocontrol Science and Technology. 2016;**26** (3):305-319

[4] Perry RN, Ehlers RU, Glazer I. A realistic appraisal of methods to enhance desiccation tolerance of entomopathogenic nematodes. Journal of Nematology. 2012;**44**(2):185-190

[5] Silver S, Dunlop D, Grove D. Granular formulation of entities with improved storage stability. WO 95/0577 (Patent). 1995.

[6] Cruz-Martínez H, Ruiz-Vega J, Matadamas-Ortiz PT, Cortés-Martínez CI, Rosas-Díaz J. Formulation of entomopathogenic nematodes for crop pest control: A review. Plant Protection Science. 2017;**53**(1):15-24

[7] Crank J. The Mathematics of Diffusion. 2nd ed. Oxford: Clarendon Press; 1975. p. 414

[8] Behle R, Birthisel T. Formulations of entomopathogens as bioinsecticides. In: Morales-Ramos JA, Guadalupe-Rojas M, Shapiro-Ilan DI, editors. Mass Production of Beneficial Organisms: Invertebrates and Entomopathogens. 1st ed. Amsterdam: Elsevier BV; 2014. pp. 483-517

[9] Matadamas-Ortiz PT, Ruiz-Vega J, Vazquez-Feijoo JA, Cruz-Martínez H, Cortés-Martínez CI. Mechanical production of pellets for the application of entomopathogenic nematodes: Factors that determine survival time of *Steinernema glaseri*. Biocontrol Science and Technology. 2014;**24**(2):145-157

[10] Patel MN, Perry RN, Wright DJ. Desiccation survival and water contents of entomopathogenic nematodes, *Steinernema* spp. (Rhabditida: Steinernematidae). International Journal of Parasitology. 1997;**27**(1):61-70

[11] Nimkingrat P, Uhlmann F, Strauch O, Ehlers RU. Desiccation tolerance of dauers of entomopathogenic nematodes of the genus *Steinernema*. Nematology. 2013;**15**(4): 451-458

[12] Wharton DA. Life at the limits: Organisms in extreme environments. 1st ed. Cambridge: Cambridge University Press; 2002. p. 307

[13] Grewal PS. Enhanced ambient storage stability of an entomopathogenic nematode through anhydrobiosis. Pest Management Science. 2000;**56**(5):401-406

[14] Bazardeh ME, Esmaiili M. Sorption isotherm and state diagram in evaluating storage stability for sultana raisins. Journal of Stored Products Research. 2014;**59**:140-145

[15] Grewal PS. Formulation and application technology. In: Gaugler R, editor. Entomopathogenic Nematology. 1st ed. Wallingford: CABI; 2002. pp. 265-287

[16] Burger CA, Shackelford CD. Evaluating dual porosity of pelletized diatomaceous earth using bimodal soil-water characteristic curve functions. Canadian Geotechnical Journal. 2001;**38**(1):53-66

[17] Khaled AR, Vafai K. The role of porous media in modeling flow and heat transfer in biological tissues. International Journal of Heat and Mass Transfer. 2003;**46**(26): 4989-5003

[18] Shackelford CD, Moore SM. Fickian diffusion of radionuclides for engineered containment barriers: Diffusion coefficients, porosities, and complicating issues. Engineering Geology. 2013;**152**(1):133-147

[19] Su Y, Davidson JH. Introduction of fluid flow and heat transfer in porous media. In: Su Y, Davidson JH, editors. Modeling Approaches to Natural Convection in Porous Media. 1st ed. New York: Springer International Publishing; 2015. pp. 1-8

[20] Lehmann P, Assouline S, Or D. Characteristic lengths affecting evaporative drying porous media. Physical Review E. 2008;**77**(5):056309

[21] Yiotis AG, Tsimpanogiannis IN, Stubos AK, Yortsos YC. Coupling between external and internal mass transfer during drying of porous media. Water Resources Research. 2007;**43** (6):W06403

[22] Shokri N, Lehmann P, Or D. Lquid-phase continuity and solute concentration dynamics during evaporation from porous media: Pore-scale processes near vaporization surface. Physical Review E. 2010;**81**(4):046308

[23] Shahraeeni E, Lehmann P, Or D. Coupling of evaporative fluxes from drying porous surfaces with air boundary layer: Characteristics of evaporation from discrete pores. Water Resources Research. 2012;**48**(9):W09525

[24] Dincer I, Hussain MM. Development of a new Biot number and lag factor correlation for drying applications. International Journal of Heat and Mass Transfer. 2004;**47**(4):653-658

[25] Stakić M, Cvetinović D, Škobalj P, Spasojević V. An initial study on feasible treatment of Serbian lignite through utilization of low-rank coal upgrading technologies. Chemical Engineering Research and Design. 2014;**92**(11):2383-2395

[26] Brutsaert W. Evaporation into the Atmosphere: Theory, History and Applications. 1st ed. Dordrecht: Springer Netherlands; 1982. p. 299

[27] Kröger DG, Branfield GR. Evaporation from a water surface: Theory and experiment. R & D Journal. 2007;**23**:5-11

[28] Jodat A. An experimental study of the ability of similarity theory to predict water evaporation rate for different convection regimes. International Journal of Multidisciplinary Sciences and Engineering. 2012;**3**(2):20-30

[29] Shokri N, Lehmann P, Or D. Characteristics of evaporation from partially wettable porous media. Water Resources Research. 2009;**45**(2):W02415

[30] Yiotis AG, Salin D, Yortsos YC. Pore network modeling of drying processes in macroporous materials: Effects of gravity, mass boundary layer and pore microstructure. Transport in Porous Media. 2015;**110**(2):175-196

[31] Bergman TL, Lavive AS, Incropera FP, Dewitt DP. Introduction to Heat Transfer. 6th ed. New Jersey: John Wiley & Sons; 2011. p. 960

[32] Prakash M, Turan OF, Li Y, Thorpe GR. CFD modelling of natural convection heat and mass transfer in hygroscopic porous media. Drying Technology. 2000;**18**(10):2175-2201

[33] Chemkhi S, Khalfaoui K, Zagrouba F. Modelling of saturated porous media drying: Heat and mass transfer coupled with the material mechanical behaviour. International Journal of Chemical Engineering and Applied Sciences. 2013;**3**(1):1-6

[34] Khalfaoui K, Chemkhi S, Zagrouba F. Modeling and stress analysis during drying of a deformable and saturated porous medium. Drying Technology. 2013;**31**(10):1124-1137

[35] Manel BA, Mihoubi D, Jalila S, Ahmed B. Strain–stress formation during stationary and intermittent drying of deformable media. Drying Technology. 2014;**32**(10):1245-1255

[36] Sam Saguy I, Marabi A, Wallach R. New approach to model rehydration of dry food particulates utilizing principles of liquid transport in porous media. Trends in Food Science & Technology. 2005;**16**(11):495-506

[37] Derossi A, Severini C, Cassi D. Mass transfer mechanisms during dehydration of vegetable food: Traditional and innovative approaches. In: El-Amin M, editor. Advanced Topics in Mass Transfer. 1st ed. Rijeka: InTech; 2011. pp. 305-354

[38] Coussot P. Scaling approach of the convective drying of a porous medium. The European Physical Journal B. 2000;**15**(3):557-566

[39] Prat M, Agaësse T. Thin porous media. In: Vafai K, editor. Handbook of Porous Media. 3rd ed. Boca Raton: CRC Press; 2015. pp. 89-112

[40] Prat M. Recent advances in pore-scale models for drying of porous media. Chemical Engineering Journal. 2002;**86**(1-2):153-164

[41] Mortier STF, De Beer T, Gernaey KVRemon JP, Vervaet C, Nopens I. Mechanistic modelling of fluidized bed drying processes of wet porous granules: A review. European Journal of Pharmaceutics and Biopharmaceutics. 2011;**79**(2):205-225

[42] Litster J, Ennis B. The Science and Engineering of Granulation Processes. 1st ed. Dordrecht: Springer Netherlands; 2004. p. 250

[43] Torquato S. Optimal design of heterogeneous materials. Annual Review of Materials Research. 2010;**40**(1):101-129

[44] Metzger Y, Tsotsas E. Network models for capillary porous media: application to drying technology. Chemie Ingenieur Technik. 2010;**82**(6):869-879

[45] Caccavale P, De Bonis MV, Ruocco G. Conjugate heat and mass transfer in drying: A modeling review. Journal of Food Engineering. 2016;**176**:28-35

[46] Kechaou N, Maâlej M. A simplified model for determination of moisture diffusivity of date from experimental drying curves. Drying Technology. 2000;**18**(4-5):1109-1125

[47] Kowalski SJ, Rybicki A, Rajewska K. Optimal control of convective drying of saturated porous materials. AIChE Journal. 2013;**59**(12):4846-4857

[48] Clarke A, Blake TD, Carruthers K, Woodward A. Spreading and imbibition of liquid droplets on porous surfaces. Langmuir. 2002;**18**(8):2980-2984

[49] Báez-González JG, Pérez-Alonso C, Beristain CI, Vernon-Carter EJ, Vizcarra-Mendoza MG. Effective moisture diffusivity in biopolymer drops by regular regime theory. Food Hydrocolloids. 2004;**18**(2): 325-333

[50] Lyn ME, Burnett D, Garcia AR, Gray R. Interaction of water with three granular biopesticide formulations. Journal of Agricultural and Food Chemistry. 2010;**58**(3):1804-1814

[51] Kelly-Zion PL, Pursell CJ, Vaidya S, Batra J. Evaporation of sessile drops under combined diffusion and natural convection. Colloids and Surfaces A: Physicochemical and Engineering Aspects. 2011;**381**(1-3):31-36

[52] Pan Z, Dash S, Weibel JA, Garimella SV. Assessment of water droplet evaporation mechanisms on hydrophobic and superhydrophobic substrates. Langmuir. 2013;**29**(51):15831-15841

[53] Pan Z, Weibel JA, Garimella SV. Influence of surface wettability on transport mechanisms governing water droplet evaporation. Langmuir. 2014;**30**(32):9726-9730

[54] Kovalchuk NM, Trybala A, Starov VM. Evaporation of sessile droplets. Current Opinion in Colloid & Interface Science. 2014;**19**(4):336-342

[55] Rahimi A, Metzger T, Kharaghani A, Tsotsas E. Interaction of droplets with porous structures: Pore network simulation of wetting and drying. Drying Technology. 2016;**34** (9):1129-1140

[56] Liu M, Wu J, Gan Y, Hanaor DAH, Chen CQ. Evaporation limited radial capillary penetration in porous media. Langmuir. 2016;**32**(8):9899-9904

[57] Tong CH, Lund DB. Effective moisture diffusivity in porous materials as a function of temperature and moisture content. Biotechnology Progress. 1990;**6**(1):67-75

[58] Popiel I, Holtemann KD, Glazer I, Womersley C. Commercial storage and shipment of entomogenous nematodes. US5183950 A (Patent). 1993

[59] Bedding RA, Wang JX, Bulter KL. Method for packaging entomopathogenic nematodes for storage and transport. WO 94/19940 (Patent). 1994